C++ Programming
Fundamentals

C++ Programming Fundamentals

Chuck Easttom

Charles River Media, Inc.
Hingham, Massachusetts

Publisher: David Pallai
Production: Paw Print Media
Cover Design: The Printed Image

CHARLES RIVER MEDIA, INC.
20 Downer Avenue, Suite 3
Hingham, Massachusetts 02043
781-740-0400
781-740-8816 (FAX)
info@charlesriver.com
www.charlesriver.com

This book is printed on acid-free paper.

Chuck Easttom. *C++ Programming Fundamentals.*
ISBN: 1-58450-237-1

Library of Congress Cataloging-in-Publication Data

Easttom, Chuck.
  C++ programming fundamentals / Chuck Easttom.
      p. cm.
Summary: Introduces basic concepts of C++ programming, including Microsoft Visual C++, using examples from such areas as game programming and GPA calculation, and provides fully functional sample programs.
    ISBN 1-58450-237-1 (paperback with CD-ROM : alk. paper)
  1. C++ (Computer program language)—Juvenile literature. [1. C++ (Computer program language)
2. Programming (Computers) 3. Computers.] I. Title.
  QA76.73.C153 E23 2003

                        2002151914

Printed in the United States of America
02 7 6 5 4 3 2 First Edition

Contents

· ·

Acknowledgments

No book is really the work of a single person, even if only one name appears on the title. First and foremost, I would like to thank a few of my students and one of my colleagues that reviewed rough drafts for me. Patrick Langlinais, Nicholas Russo, and Susan Hebert were all gracious enough to look over rough drafts for me and to give me their opinions and their encouragement. The publishing team at Charles River Media has also been simply amazing. They carefully edited and reviewed the rough drafts, saving me from embarrassing myself! Finally, I must also thank my wife, Misty, and my son, AJ. Without their patience and support, none of the books I write would be possible.

Introduction

This book is about C++, a fact you are undoubtedly already aware of. There are a lot of other C++ books out there. That's another fact you are probably already aware of. Why this one? What is different? First of all, this book is aimed at the beginner, but does not talk down to the reader. Yes, you may be a beginner at C++ and perhaps even a beginner at programming. But that does not mean you are intellectually challenged, and this book makes no such assumptions. Also this book accepts that as a beginner, code snippets are totally inadequate. You need complete and total code samples. Samples that you can run exactly as they are written. For this reason, this book has over 80 completely working code samples, completely written out in the text, and on the CD-ROM. There are still a few code snippets scattered throughout the text, but only to illustrate key techniques that are then demonstrated in a complete working program.

ON THE CD

In this book, we will step through the various parts of C++ programming in a very systematic way. Many of the concepts you will learn are essential programming concepts, simply applied using the C++ language. Along the way the book also strives to teach you ANSI-standard C++. However, occasionally the text does add-in commonly used techniques that are not part of the formal ANSI standard. These items will be noted when they appear. Finally, the book gives you a peek at *Visual C++* programming. This is done because Microsoft *Windows* programming is just so common, it would be a serious omission if the text did not introduce you to this topic. Essentially what this book proposes to do is to teach you the fundamentals of computer programming, via the C++ programming language.

This book endeavors to present C++ in a clear and understandable way without talking down to you. The first few chapters use rather standard, simple, example code. However, as soon as you have gotten enough knowledge to move on to more interesting code, the book starts using real-world, interesting examples. Farther along in the book you are even shown how to do a simple game. However, the bulk of this book is concerned with teaching the fundamental programming concepts of C++.

A few notes about this book's coding and writing style. To begin with, this book is written in a manner that is easy for beginners to follow. This means that it is not always the most compact code. A lot of C++ programmers write their code in the most compact fashion possible. And that is fine, it's just difficult for some beginners to follow. So, the code in this book is presented in a manner that is easiest for a beginner to follow. The writing style includes a habit of giving a lot of examples. The thought is that someone learning a new programming language should have a lot of examples to look at.

C++ Fundamentals

The purpose of this section is simply to lay down the foundations of C++. These are the core concepts in programming and in C++ that you will need throughout this book. Although the examples in this section are not as exciting as those in later sections, they are very important. If you do not master the topics in this section, you will be unable to master the rest of this book. At the end of this section, you should be able to write basic C++ programs. These first nine chapters represent the core of C++. You cannot even begin to consider yourself a programmer, much less a C++ programmer, if you do not master these first eight chapters.

1

C++ Basics

IN THIS CHAPTER
· · · · · · · · · · · · · ·

- History of C++
- What Is C++?
- How to Write a C++ Program
- C++ Fundamentals
- Statements and Expressions
- Basic Structure of a C++ Program
- Function Basics
- Variable Scope
- Compiling
- Commenting Your Code

Welcome to C++! This first chapter will provide you with some basic foundational material you will need to progress through the rest of the book. In this chapter you will be introduced to the history of C++, the basics of the language, and how to use variables and write expressions in C++. These fundamental concepts are the essential building blocks that you will use to create C++ programs throughout the rest of this book.

HISTORY OF C++

C++, as the name implies, is essentially based on the C programming language. Therefore, it seems prudent to begin with a brief history of C. The C programming language was devised in the early 1970s at Bell Laboratories by Dennis Ritchie. It was designed as a system implementation language for the Unix operating system. The history of C and Unix are closely intertwined. For this reason a lot of Unix programming is done with C. To some extent, C was originally based on the typeless language BCPL, however it grew well beyond that.

The C++ programming language was invented by Bjarne Stroustroup. Work on what would become C++ began in 1979. The initial version was called "C with Classes." That name did not work out well, and was replaced with C++. The first version of C++ was used internally in AT&T in August 1983. The first commercial implementation was released in 1985. The C++ language standards are now handled by the American National Standards Institute (ANSI), and the International Standards Organization (ISO). This is why you often hear pure C++ referred to as ANSI Standard C++, or ISO Standard C++.

HINT!

Pure C++ is mentioned because there are a lot of extensions that are specific to a particular compiler or operating system. A few of these are covered in this book, but are identified as being nonstandard.

WHAT IS C++?

You can see that the C programming language was developed first, C++ was developed later. You might be asking yourself what, exactly, is C++ and how does it relate to C? The answer is that C++ is essentially C taken to the next level. The most obvious difference between the two is that C++ supports object orientation (more on that in Chapters 10, 11, and 12). However, C++ sports many other improvements over C. For example, C++ handles strings better than C, and has a more robust exception handling. (Exception handling refers to a program's ability to handle unexpected errors. What if the user inputs a zero then tries to divide by that number? This is an exception, how your code handles it is exception handling. Chapter 7 discusses this topic in depth.)

C code will compile fine in most C++ compilers, but the reverse is not true. C++ code will not necessarily compile in a C compiler. You may be wondering what is meant by the word *code*. Code is essentially the series of programming commands that a programmer writes. All the commands that make up a program are the source code for that program.

C++ supports all C commands and also has many additions. You may frequently see old style C code mixed in with C++ code, especially in programs written by programmers who originally started in C.

HOW TO WRITE A C++ PROGRAM

There are several ways to write a C++ program. You could simply open your favorite text editor, such as *notepad*, write your program, save it, and then use a command line compiler to compile the program. In fact, this is the method that will be used in this book, until we get to Section IV, the section on *Visual C++*. If you have a commercial development tool such as Borland *C++ Builder*™ or Microsoft *Visual C++*,™then you must follow the instructions for that particular software package. The examples in this book will use the free downloadable version of the Borland C++ compiler. The free Borland C++ compiler does not have the extra development tools and IDE (Integrated Development Environment) that Borland *C++ Builder* has. The Web address to go to and download this free C++ compiler is listed in Appendix A. However, the free download is simply a command line compiler. A command line compiler is one that is executed from the command line. This, of course, begs the question of what is a command line. When you enter commands at a DOS prompt (Microsoft *Windows*™98), Command Prompt (Microsoft *Windows 2000/XP*), or Shell (*Linux/Unix*™) you are typing commands on a line, thus the term "command line." You literally type in commands one line at a time. Figure 1.1 shows the basic DOS prompt for *Windows 95/98*™.

To use a command line compiler, like the free C++ compiler from Borland, you simply type the code into a document in any text editor, then save it with a .cpp extension (e.g., myprogram.cpp). That .cpp extension stands for C Plus Plus. C files that are done in a text editor are saved with the extension .c, as you might expect. Complete instructions for how to download, install, configure, and use this compiler can be found in Appendix D. You should use the instructions in Appendix D to make

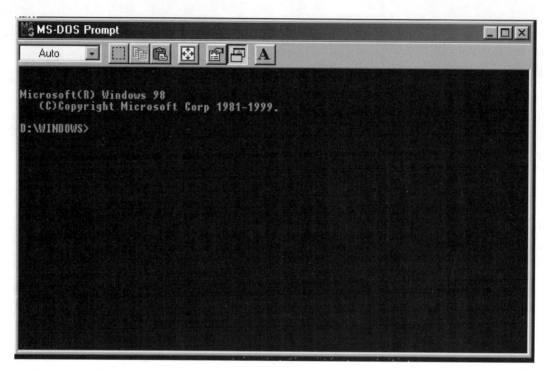

FIGURE 1.1   DOS Prompt.

sure you have the compiler properly installed on your PC before continuing with this book.

C++ FUNDAMENTALS

C++ is a programming language. As such, it shares a number of similarities with other programming languages. First we may need to clarify what a programming language is. A computer thinks in 1's and 0's. Various switches are either on (1's), or off (0's). Most humans, however, have trouble thinking in 1's and 0's. A *programming language* is a language that provides a bridge between the computer and human beings. A "low-level" language is one that is closer to a computer's thinking than to a human language. A prime example would be Assembly language. A "high-level" language is one that is closer to human language. COBOL and BASIC are prime examples of high-level language. Many people consider C and C++ to be high-level languages, but actually C and C++ are both somewhat of a bridge between the low-level languages and high-level languages. You might think of them as *mid-level languages*. The level of a language, in this context, essentially refers to how far it is removed from actual machine language. It has no relationship to either the power of the programming language, or the difficulty of learning and using it. Various languages, with their basic *level* indicated, are shown in Figure 1.2.

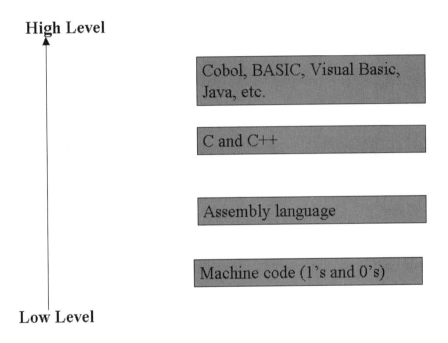

FIGURE 1.2   Programming languages.

Each programming language has various strengths and weaknesses. Some, like BASIC, are easy to use but are neither flexible nor powerful. Others, such as Assembly, are powerful, but difficult to use. C and C++ are somewhere in the middle. C++ is quite powerful, but relatively easy to use (easier than Assembly but more difficult than BASIC). Some languages are also written with specific purposes in mind. Fortran was written specifically for mathematical programming, BASIC was designed simply for teaching programming, and COBOL was designed for business applications. C++ was designed as a general purpose language. It is used in business applications, telecommunications programming, artificial intelligence programming, games programming, and much more. This is one reason why it is the language that many teachers choose to teach students. Once you have learned C++, you can use it in a wide variety of situations.

Programs are written to handle data. This is why the industry as a whole is often referred to as data processing, information technology, computer information systems, and so on. That data might be information about employees, parts to a mathematical computation, scientific data, or even the elements of a game. No matter what programming language or techniques you use, the ultimate goal of programming is to store, manipulate, and retrieve data. Data must be temporarily stored in the program, in order to be manipulated. This is accomplished via *variables*. A *variable* is simply a place in memory set aside to hold data of a particular type. It is a specific section of the computer's memory that has been reserved to hold some data. It is called a variable because its value or content can vary. When you create a variable, you are actually setting aside a small piece of memory for storage purposes. The name you give the variable is actually a label for that address in memory. For example, you might declare a variable in the following manner.

```
int j;
```

Then, you have just allocated four bytes of memory; you are using the variable j to refer to those four bytes of memory. You are also stating that the only type of data that j will hold, is whole numbers. (The int is a data type that refers to integers, or whole numbers.) Now, whenever you reference j in your code, you are actually referencing the contents being stored at a specific address in memory.

With that said, you might now be asking what is meant by "data of a particular type"? Data comes in different types. Some data consists of letters and some consists of numbers. Any programming language will recognize only certain types or categories of data. The basic data types that C++ recognizes, and what they hold, are shown in Table 1.1. In addition, there are a few other data types that will be introduced later in the book.

When you are deciding what type of variable to create, you need to think about what kind of data you might wish to store in that variable. If you want to store a person's age, then an int is a good choice. The int data type will hold whole numbers, and will, in fact, hold numbers much larger than you might need for a person's age. The long data type will

TABLE 1.1   Data Types

Data Types	Values stored
int	This value is for whole numbers, or integers. The size depends on the operating systems. In 32-bit operating systems, the int is usually 32 bits (4 bytes) in length.
long	This data type holds larger whole numbers
float	Floats are used to hold decimal numbers such as 2.02798
double	A double is simply a really big float.
char	A char can store a single alpha-numeric type. In other words, it can store a number or a letter.
bool	Bool is short for Boolean, a reference to Boolean logic. Therefore, this data type can only store true or false.
short	This is basically a small integer. Usually 2 bytes in length. But the actually size will depend on the operating system. On 32-bit operating systems such as like *Windows 98, 2000, NT,* and *XP,* a short will be 2 bytes long.

also hold whole numbers, but it is designed for even larger whole numbers. If you wished to store bank balances, grade point averages, or perhaps temperatures, you would need to consider using a float. The float data type holds decimal values. However if you were storing astronomical data, you would still want to hold decimals, but because they might be quite large, you would have to consider storing them in doubles. Picking what data type to use is actually simple. Just give some thought to the type of data you intend to store in a variable.

In addition to considering what type of variable you will use, you must also pay attention to how you name the variable. Only certain names are allowed in C++. Variable names must begin with a letter or underscore and may contain any combination of upper/lower case characters, digits, and underscores. Variable names can neither begin with a number nor contain certain symbols such as #, &, *, and so on. Here are some examples of valid and invalid variable names (Table 1.2).

In addition, many of the symbols that C++ does not allow you to use in naming your variables actually represent something important in the C++ programming language. This is one reason why such symbols are not allowed in variable names. Also, please understand that simply because a name is valid in C++ does not make it a good name. It is

TABLE 1.2   Valid Variable Names

Valid Variable Names	Invalid Variable Names
Accountnumber	&account
_LastName	Last:Name
Length_of_side	#length
Temperature	$temp

important that your variable names bear some resemblance to the value they hold. The compiler will certainly let you compile your code without this, but it will make your code much more difficult for other programmers to understand. If your variable holds an account number, please don't name the variable x or i. Instead, give it a name such as account_number or acctnum.

Some programmers go further and add a notation at the beginning of the variable name to indicate what type of variable it is. For example, an account number that is an int might be called i_accountnum or intAccountNum. The following table shows common prefixes used for various data types. Not everyone follows these naming conventions. Nevertheless, they are common enough that you should be aware of them. They are summarized in Table 1.3.

There is a plethora of ways to name variables; no way is right or wrong. The thing to keep in mind is this: Does your personal naming convention make it easy for others to read your code? Does your variable's name clearly identify what type of variable it is and/or what data it will hold? If you can answer *yes,* then your naming convention is fine.

TABLE 1.3   Naming Conventions

Variable Type	Common Naming Conventions
int	iaccountnumber, i_accountnum, intaccountnum
float	fbalance, f_balance, flt_balance
long	lnumber, l_number, lngnumber
bool	b_isempty, bool_isempty
char	c_value, chrValue

Now that you know how to name variables, let's give you a few examples.

```
int int_account_num;
boolean bturnedon;
float fltsalary;
```

Notice that these examples all obey the aforementioned rules for naming variables. You might, however, be curious about the semicolon at the end of each line. Each statement in C++ (as well as C, Sun *Java*, and several other programming languages) ends with a semicolon. The semicolon basically tells the compiler that you are done with that particular line of code, and that the things you have written prior to that semicolon represent one single, concise statement. A *statement* is simply a single line of code that performs some action. Many programmers also use the term *expression* when referring to a statement. The two terms, expression and statement, are interchangeable.

You can declare more than one variable on a single line. All the variables declared in that statement will be of the type you declared at the beginning of the statement.

```
int i_account_num, i_age, int_years_w_company;
```

All three variables are of type `int`.

HINT!

This works the same way in many other programming languages such as *Java* and C, but does NOT work this way in *Visual Basic™*.

WATCHOUT!

C++ is case sensitive. Uppercase letters and lowercase letters are treated as two different letters. An *A* is not the same as an *a*. That means that `int main()` and `int Main()` are not the same thing. One of the most common mistakes beginners make is forgetting the case-sensitive nature of C++.

You can choose to initialize your variable to some default value when you create it. The following examples illustrate this.

```
int num = 0;
float number = 0.0;
```

A default value is simply some starting value that the variable will hold, by default, if no other value is placed into it. For example, if your program was storing data about people who recently received a high school diploma, you might wish to use a default value of 18, because that is the most probable age of a recent high school graduate.

STATEMENTS AND EXPRESSIONS

As you have already seen, a *statement* is simply a single line of code that performs some action. Remember that another word for a statement is an *expression*. In C++, a single expression does some type of action. That action might be to declare a variable, add two numbers, compare two values, or just about anything at all. It is also important to remember that all statements/expressions end with a semicolon. Another of the most common mistakes that beginners make is leaving off semicolons. Let's look at a few simple statements.

```
int i_acctnum;
i_acctnum = 555555;
i_acctnum = i_acctnum + 5;
```

Each statement performs a different action—but it does perform some action—and it ends with a semicolon. The first statement/expression simply declares a variable. The second statement/expression sets that variable equal to some value. Finally, the last statement/expression performs addition and puts the answer in the variable that was previously declared.

Operators

You saw, at the end of the last section, the use of the + sign. This is an *operator*. An *operator* is simply some symbol that performs some action, usually a mathematical action, such as addition or subtraction. C++ supports a number of important operators that you will need to get familiar with. Let's begin examining C++ operators, starting with the basic math operators, shown in Table 1.4.

TABLE 1.4   Operators

Operator	Purpose	Example
+	To add two numbers	int answer; answer = 5 +6;
-	To subtract two numbers	int answer; answer = 10 –3;
*	To multiply two numbers	int answer; answer = 4 * 5;
/	To divide two numbers	int answer; answer = 7/3;
++	This is a special operator that simply increments the value by 1. You will see this used later in this book when loops are discussed.	int answer; answer++;
--	This is also a special operator that simply decrements the value by 1.	int answer; answer--;
=	The single equals sign is an assignment operator. It states "make what's on the left equal to what's on the right"	answer = 16;
==	The double equals is an equality operator. It asks "is what's on the left equal to what's on the right?" This is frequently used in if statements (which you will see in a later chapter!)	if(answer==5)
!=	Not equal to	if (x !=3)
+=	Add then assign	x += 1;
-=	Subtract then assign	x -=1;
||	This is the logical OR operator.	if(j == 5 || j ==10)
&&	This is the logical AND operator	if(j > 5 && j<10)
>>	Bitwise shift to the right	3<<2;
<<	Bitwise shift to the left	3>>2;
&	Bitwise And	3&2;
|	Bitwise Or	3|2

WATCHOUT!

 When using the increment (++) or decrement (—) operator, it is necessary to be careful of where you put it. If you put the operator before the variable, then that operation will take place before any other operations in the expression. Some examples follow.

```
int answer, num;
num = 5;
```

Now using the expression,

```
answer = num++;
```

the value of num will first be put in answer, THEN incremented. In other words, the value of answer will be set to 5, and then the value of num will be incremented. If you want num to be incremented BEFORE you assign the value to answer then you must put the increment operator first, as shown in the following example.

```
answer = ++num;
```

The operators shown in Table 1.4 are your basic math operators and logical operators. There are several other operators you will also see in C++. These will be introduced in later chapters because they are pertinent to topics that will be covered in those chapters. Most of the operators in Table 1.4 should be familiar to you. The math operators represent simple math operations. All math operators occur with a specific order. The order is pretty much the same as the order of operations defined in mathematics. First is * (multiplication), next / (division), then + (addition), and, finally, – (subtraction). The increment ++ and decrement – operators precedence is determined by what side of the variable they are on. You can alter the order of operations by using parentheses, just as you do in mathematics. An example follows.

```
answer = x * 3 + 4;
```

First, x and 3 would be multiplied, and then 4 would be added to that product. If, instead, you wanted to add 3 and 4, then multiply by x, you could simply use parentheses to denote that.

```
answer = x * (3 +4);
```

The order of multiplication-division-addition-subtraction is simply the order defined by mathematics for these operations. Some teachers even use the mnemonic *My Dear Aunt Sally* to help students to recall the proper order of operations. The important thing to remember is that an operator is simply a symbol that defines some action to be taken. The + symbol defines the action of addition and the \ symbol defines the action of division. Other operators, such as the increment ++ operator, may be new to you, but the fact remains that they are simply symbols that define some action.

An operator is considered to be a *unary* operator if it takes only one argument. Put another way, if a symbol only acts on one number then it is a unary operator. The increment and decrement operators fall into this category. They only work on a single variable. Binary operators, on the other hand, work on two numbers. Addition and subtraction both work on two numbers. A more technical definition would be that *binary* operators are operators that take two arguments.

BASIC STRUCTURE OF A C++ PROGRAM

Now that you have seen variable declaration, C++ expressions, and the basic math operators, let's take a quick look at the basic structure of a C++ program. All C++ programs must have a main function. This is where the program begins. Chapter 4 goes into greater detail regarding functions. For now, suffice it to say that a *function* is one or more statements grouped together in a logical manner to accomplish some goal under a common name. Chapter 4 covers all the intricacies of C++ functions. The following is a basic C++ program.

```
int main()
{
    return 0;
}
```

Now, this program will not do much—in fact, it won't do anything useful at all. It is, however, a valid C++ program. It has a main function that returns an integer value. The main function is where all C++ programs start. This is the starting point for your entire program. All main functions are required by the ANSI (American National Standards Institute) standards to return an integer. The integer is a 0 if the program executes fine, and a 1 if some problem is encountered. This was originally done because some operating systems require any program to return a value telling the operating system that everything is OK. You will see a lot of *Windows* programmers use a main function that returns a void like the following.

```
void main()
```

This will work in some cases, but it is not technically correct. This book endeavors to conform to the ANSI standards and use return 0.

The next thing to notice about our sample program is the presence of brackets. These brackets are C++'s way of establishing borders around any block of code. (Incidentally, C, *Java*, and *JavaScript*™ do the same thing. Learning it here will help you learn other languages as well.). Every time you see an opening bracket, {there must be a matching} closing bracket. You will see a lot more about this when we discuss loops, decision structures, and functions. For now, suffice it to say that brackets form borders around blocks of code.

With all that said, let's look at a program that has a few statements.

```
int main()
{
 int j;
 j = 5;
 j++;
 return 0;
}
```

This simple program creates a variable named j, sets the value of that variable to 5, then increments that value by one. This is still not a particularly exciting piece of code, but it does illustrate the basic structure of a C++ program. Three statements are executed, a 0 is returned to indicate that everything is OK, and there are brackets surrounding the main function. If you carefully examine this code, and follow this template in all your programming, then you will do well!

Header Files

There is only one more item that needs to be discussed regarding the basic structure of a C++ program. That item is header files. *Header files* are files that contain the definitions for functions and/or classes. Basically, if you have a lot of functions or variables that you might want to use in several programs, you can put them in a header file, then include that header file anywhere you need to use it. After creating a header file you can then include a reference to a header file in your program and use the functions and/or classes defined in that header file. The real intricacies of header files are discussed in a later chapter, in which you will actually create some header files. However, you should realize that there are a lot of these files built into C++ that you can use! Well, "built in" is not exactly accurate. They are actually installed with your compiler and you can include them in your programs. One of which you will use in Chapter 2 is the `iostream` header file. It gives you functions to handle input and output from the screen. The way you include any header file is simple. At the beginning of your file, you reference the header file. This is demonstrated in the following example.

```
#include <iostream>
int main()
{
 return 0;
}
```

You now have access to all the classes and functions defined in the `iostream` header file. In the following chapters, you will see several different header files, each providing a different set of functions for you to use. For now, all that is important is that you realize that you include a header file using the # sign and the word *include*. If the header file is one that is part of C++, then you simply use the `<headerfilename>`. Make sure that this is the first thing you do in your file! Remember that you use .cpp to indicate a C++ source file and .c to indicate a C source file, while the .h indicates a header source file. A *source file* is simply a plain text file that has the source code for your C++ program. Source code is the basic lines of programming language that you write, and that the compiler will later use as a source to create an executable program.

HINT!
• •

C requires you to specify the .h in the file. For example, screen input and output is implemented in C using the `<stdio.h>` header file. The .h extension is not needed in C++. Many compilers will support it, but not all. It is best to refer to `<iostream>` rather than `<iostream.h>`. It is also important to realize that the ISO and ANSI C++ standard dictates that, unless you are using an old C-styled header file, you will not add the .h extension.

FUNCTION BASICS

What are functions? In simple terms, a *function* is a block of code that is identified by some name and can take zero or more parameters and can return some value. That definition probably does little to clarify the issue for you! Let's look at a function and its parts to help clarify this issue. To begin with, consider the following function, which squares a number.

```
float square(float num)
{
  float answer = num * num;
  return answer;
}
```

The first line of the function is called its declaration line. It has three parts. The first part is its return type. The return type tells you what kind of data the function will give you back. If it returns nothing, then its return type would be void. In the previous example, the function will return the answer to our math problem, and that is a float. The second thing is the name of the function. Function names follow the same guidelines as variable names. It is a good idea to have the function name reflect what the function will do. Next, inside the parenthesis, we have parameters. A parameter is some value you pass to the function in order for it to work. For example, in this case, to square a number, we have to give it the number to square. Students often ask: "What do we need to put as parameters? Do we even need any parameters for this function?" There is a simple way to answer this question for any function you write. Ask yourself this question: If you wanted some person to do this task for you, would you need to give them anything? And, if so, what? If you wanted a person to square a number for you, you would have to give them that number. However, if you just wanted them to say hello, you would not have to give them anything. Therefore, a function that squares a number should take one parameter, and a function that displays hello on the screen might not take any parameters. There will be a lot more in-depth discussion on functions in Chapter 4.

VARIABLE SCOPE

In the beginning of this chapter, you were introduced to the concept of variables. You saw the various data types, how to declare a variable, and what proper names you could use for variables (as well as ones you could not). There is one more topic concerning variables. This had to be left until after the discussion about the structure of a C++ program, and you will see why shortly. The topic at hand concerns WHERE you declare a variable. This is called variable scope. Consider the following example.

```
int main()
{
   int j;
   return 0;
}
```

This variable is declared inside the main function. That means it is only usable within that function. If you create other functions, they will not be able to access the variable j. This is referred to as a *local variable*. A local variable is a variable that is declared within a function, or within any block of code. Now let's look at another example.

```
int j;
int main()
{
    j++;
    return 0;
}
```

Notice that *j* was NOT declared inside the main function. It was declared outside of any function. This means that it can be used throughout your code. A variable declared outside of any function can be used in any function, and is referred to as a *global variable*. This is referred to as variable scope. Scope defines the range within which a variable can be used. If the variable has local scope, then its range is restricted to the function in which it was declared. It is essential to pay attention to the scope of your variables. It is not uncommon for a beginner to declare a variable inside one function, and then forget that it is only usable within that function.

COMPILING
• • • • • • • • •

When you write a program you are typing in code. You might be simply typing code into a basic text editor or into a commercial IDE (Integrated Development Editor) such as Borland C++ *Builder* or Microsoft *Visual*

C++. But you will not be distributing what you are typing to end-users. What do you distribute to them? First of all, what you type is the source code. It is not an executable program. Remember that the source code is simply all the programming commands that you write. The process of taking your source code and creating an executable is called compiling. The purpose of a compiler (whether it's a command line compiler, or one built into some commercial IDE) is to translate the specific programming language code you wrote into standard machine instructions that the computer can understand and run. That code is in a binary data file (1's and 0's) that the machine can understand and execute (thus, the term *executable*).

The specifics about how you compile a program depend on the compiler you are using. Commercially available development kits such as *C++ Builder and Visual C++* usually have a button to press, or a dropdown menu to select. Command line compilers require you to type in compilation instructions at the command line. The specifics for using the free Borland C++ compiler used in the examples in this book are given in Appendix D.

COMMENTING YOUR CODE

After you have worked out code that does some particular action, what happens when the code has to be modified 6 months later? You probably won't remember what you were thinking or intending when you wrote the code, and you will have to spend significant time just trying to remember what you were trying to do. It is even worse if you have to modify someone else's code. Well, never fear—there is an answer. C++ (as well as all other programming languages) provides a way for you to put comments in your code. A comment is simply an explanatory note that the compiler will ignore. Comments are for the programmer working on the code to read, not for the compiler. There are two ways to make a comment. The first is to put two forward slashes // at the beginning of that line, like you see in the following example.

```
// This variable is used to hold account numbers
int account_num;
```

The compiler will ignore the comment, but any programmer working on the code can read it to see what the code in question was meant to do. It is a common practice to put a few lines of comments at the beginning of a function, to explain the basic purpose of that function. Another way to make comments, if you have several lines of comment, is to enclose the comments with a beginning forward slash asterisk /* and end the comments with an asterisk forward slash */. Here is an example.

```
float square_number(float num)
{
/* This function takes in a single number then squares
   that number
 (multiplies it by itself) and returns the answer */
 return num*num;
}
```

Functions are simple. As such, some programmers skip comments all together. Your code will compile and run without comments. However, when you have to come back later and modify the code, you will be sorry if you did not comment it properly. I strongly encourage you to start early using comments in your code. Throughout the rest of this book you will frequently see comments used.

SUMMARY

This chapter is just the beginning of your journey into C++. After reading this chapter you should be comfortable with the basic structure of a C++ program, as well as variables, operators, and expressions. Make sure you carefully study this chapter, as these concepts are key to understanding the rest of this book. Also make certain that you can answer all the review questions before proceeding to the next chapter.

REVIEW QUESTIONS

1. What is a variable?
2. What is a statement?
3. Which of the following are legal names for variables in C++?
   a. firstname
   b. &salary
   c. x

       d. y

       e. first@name

4. All valid C++ statements end with a what?
5. What type of value must the main function return?
6. What is the purpose of brackets?
7. What is a local variable?
8. What is a global variable?
9. If a function does not return anything, then what do you put for its return type?
10. What is a parameter?

Console Input
and Output

The previous chapter introduced you to some basics of C++ program-ming. This chapter covers another vital part of C++ programming: getting input from the user and displaying the results. This chapter con-centrates on getting that input and providing the required output. This chapter focuses on doing this from some command line such as the *Win-dows* DOS prompt/command prompt, or a *Unix/Linux* shell. This means that you will learn how to take in what the user types, and to display results in a text format, back to the user. Obviously a program that does not take in input, or give back output, is ultimately useless, so please pay particular attention to this chapter.

OUTPUT TO THE SCREEN

You will often want to provide the user with some type of text. It might be a message to the user, a request for input, or the results of some data that the program has computed. How do you take data from your program and convert it into some representation on the screen? Luckily for you, C++ provides some functions that handle screen output for you. All you have to do is learn how to use these functions. These functions are found in one of the header files that was mentioned in Chapter 1. The particular file you need to include is the iostream file. You include this by simply placing the following line at the top of your source code file.

```
#include <iostream>
```

Recall from Chapter 1 that, when you include a header file, you have access to all the functions defined in that file. By including this file you will have access to a number of functions for input and output. The two most important functions are cout and cin. Virtually all your input and output needs (at least for keyboard input and screen output) can be handled by these two functions. The first, cout, is how you display data to the screen. The following example shows the ubiquitous "hello world" program. (It seems like every programming book uses this example, so who are we to argue with tradition?)

Example 2.1

Step 1: Enter the following code into your favorite text editor then save the file as *example02-01.cpp*.

```cpp
#include <iostream>
using std::cout;
using std::cin;
int main()
{
  // Print hello world on the screen
  cout << "Hello World";
  return 0;
}
```

Step 2: To compile this, simply type the code into a text file and save it as *hello.cpp*. Then you will run your command line compiler by typing in *bcc32 hello.cpp.* If you typed in the code properly you will then have a *hello.exe,* that you can run any time you wish.

Step 3: Run the executable you just created. When you run it, it should look similar to Figure 2.1. (This image is from the command prompt of a *Windows 2000* professional machine. A *Linux* shell would look a little bit different, but would be essentially the same concept.)

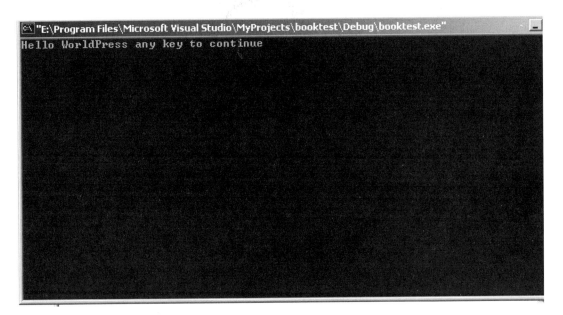

FIGURE 2.1   Hello World.

This may not be a particularly exciting program, but it does illustrate the basics of screen output. Notice that after we included <iostream> we also had two strange-looking lines of code.

```
using std::cout;
using std::cin;
```

cout and cin are both defined inside of the iostream header file. You have to tell your program which parts of iostream you wish to use. These two lines of code tell the compiler that you wish to use cout and cin. (cin will be described in detail later in this chapter.)

The cout command tells the C++ compiler to redirect the output to the default display device, usually that's the monitor. cout is short for "console output." Notice the << after the cout. The arrows literally point you in the direction the text will be sent. In the case of cout, the code is sent out of the program, thus the arrows point out of the code! This seems pretty simple so far, and it should. Now, what if we wish to format the code that we output in some special way? For example, when a program is done, the *Windows 2000* command prompt (and earlier *Windows* DOS prompts) adds on the phrase "press any key to continue." Perhaps you would like to place that on a separate line, to separate it from the output you are producing. That would be the logical thing to do because you do not wish to confuse the user into thinking that "press any key to continue" is your program's output. Fortunately for you, C++ provides several formatting codes that you can add to any string to format it. For example, the \n code tells the C++ compiler to start a new line. Let's rewrite the "hello world" program with this addition.

Example 2.2

Step 1: Type the following code into your favorite text editor and save it as *example02-02.cpp*.

```
#include <iostream>
using std::cout;
using std::cin;
int main()
{
  cout << "Hello World \n";
  return 0;
}
```

Step 2: Compile the code by running *bcc32 example02-02.cpp*.

Step 3: Execute the compiled code by typing *example02-02*. You should see an image like that depicted in Figure 2.2.

Note that anything after "Hello world" is on a new line. That's exactly what the \n command means; it means start a new line. There are actually several commands that you can execute in this manner to format the code in any way you like. These codes are often referred to as *escape codes*. Table 2.1 summarizes most of them for you.

As you can see, there is a plethora of options for formatting output to the screen. Throughout this chapter and the next you will see several of these codes used in examples. This table provides a summary of the

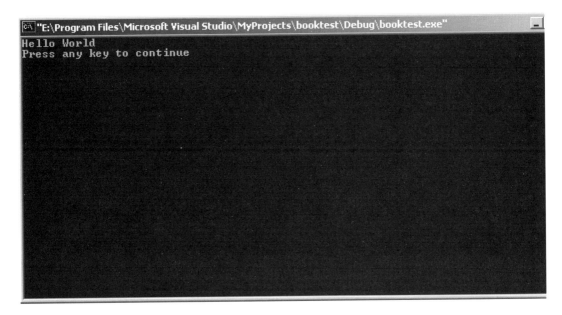

FIGURE 2.2   Hello World 2.

TABLE 2.1   Escape Codes.

Escape Sequence	Represents
\n	New line
\r	Carriage return
\b	Backspace
\t	Horizontal tab
\'	Single quotation mark
\f	Form feed
\a	Bell (alert)
\v	Vertical tab
\\	Backslash
\"	Double quotation mark
\?	Literal question mark

escape characters you can use. You should recall from Chapter 1 that C++ is built on the framework of C. These formatting keys (also called escape keys) are a prime example. These keys work exactly the same way in C as they do in C++. The following is an example that illustrates everything covered thus far.

Example 2.3

Step 1: Enter the following code into a text editor of your choice and save it as *example02-03.cpp*.

```cpp
#include <iostream>
using std::cout;
using std::cin;
int main()
{
  // the following code demo's various escape keys
  cout << "\" Hello World \" \n";
  cout << "C++ is COOL! \n \a";
  return 0;
}
```

Step 2: Compile the code by typing in *bcc32 example02-03.cpp*.
Step 3: Run your code by typing in *example02-03*.

If you entered the code in properly you will here a beep. (Remember that \a causes a beep.) You will see something similar to the image shown in Figure 2.3.

You should notice several things about this code. First, notice that you can place more than one escape sequence in order. You should notice that this was done in this example. You can use as many escape characters as is necessary in a given string of characters. You should also notice that the way to place quotes inside a string of characters is to use the proper escape character.

WATCHOUT!

If you try to simply put quotes inside of quotes, then your quotes will terminate the string. You must use the escape character. If you think about this, it makes perfect sense. As soon as the C++ compiler sees the quotation marks (if they are not preceded by the \) it will think that your string is now ending.

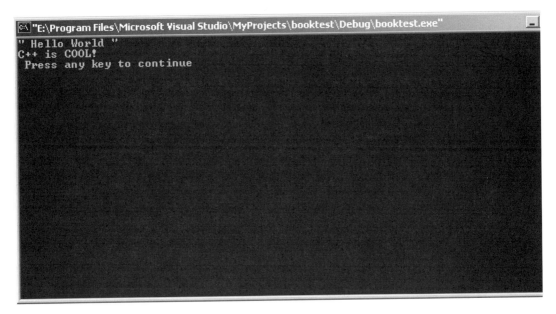

FIGURE 2.3   Hello World 3.

Finally, you should also notice the "beep" provided by \a. It is often useful to provide the user with audio signals in addition to visual signals.

Using these various escape sequences you manipulate the output of your program in a variety of ways. You can also, as you have already seen, create some audio output. The following example should illustrate this to you.

Example 2.4

Step 1: Enter the following code into your favorite text editor.

```
#include <iostream>
using std::cout;
int main()
{
cout << "As you can see these \" escape keys \" \n";
cout << "are quite \'useful \' \a \\ in your code \\ \a
    \n";
return 0;
    }
```

Step 2: Compile that code.

Step 3: Execute the code. You should see something similar to Figure 2.4.

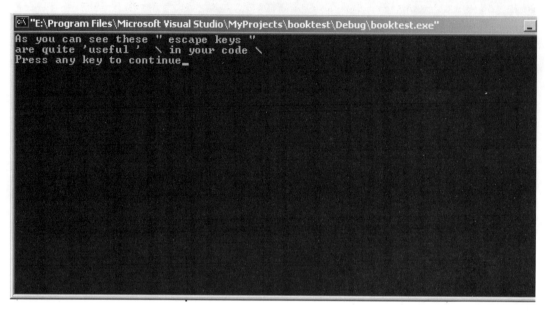

As you can see these " escape keys "
are quite 'useful '   \ in your code \
Press any key to continue_

FIGURE 2.4   Using various escape keys.

These keys give you a wide range of formatting options as well as some sound effects. You will probably find these escape keys quite useful in your various programming projects.

In addition to the escape keys you have already seen, there are some other techniques for manipulating your output. For example, you will frequently see C++ programmers choosing the endl command for a new line at the end of screen output, rather than the escape key \n. To use endl just end your quotation marks, then type the << endl command, and terminate it with a semicolon. The following example demonstrates this.

Example 2.5

Step 1: Enter the following code into your favorite text editor.

```
#include <iostream>
using std::cout;
using std::cin;
```

```
using std::endl;
int main()
{
 cout << "You have previously used the \\n key to get a
   new line \n";
 cout << "However you can also use the endl
   command"<<endl;
  return 0;
}
```

WATCHOUT!

You will notice that two of the lines of code in this sample wrap to the next line. This is simply a book formatting issue. When you type code into your text editor, put everything on one line, up to the semicolon.

Step 2: Compile that code.

Step 3: Execute the code. You should see something similar to what is displayed in Figure 2.5.

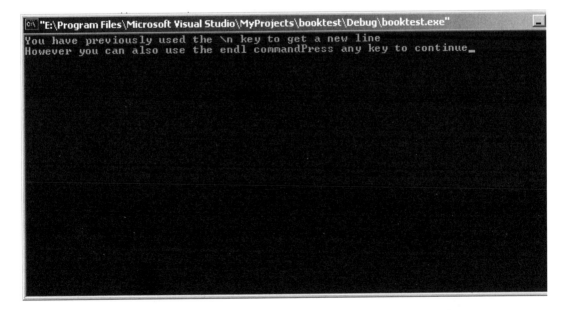

FIGURE 2.5 Using endl.

As you can see, the endl is just as useful in creating a new line. From a technical perspective, it also flushes C++'s buffer stream. What that means is that as you send content to the screen via cout, it is placed in a temporary buffer. The endl command causes that buffer to be emptied.

KEYBOARD INPUT

You can now display text to the screen. It is also a good idea to get data in from the user. Fortunately, C++ makes that just as easy as output. In fact, it is done in a similar manner. Just as output is handled with the cout command, input is handled with the cin command. There are really only two differences. The first difference is that the arrows point into the program >>. (That's logical, isn't it?) The second difference is that the arrows must point to some variable. This makes sense if you think about it. If you are going to take input from the keyboard, you have to put it somewhere. An example follows.

Example 2.6

Step 1: Type the following code into your favorite text editor.

```
#include <iostream>
using std::cout;
using std::cin;
int main()
{
  // variable declarations
  int number1, number2, product, sum;
  // get the users input
  cout<<"Please enter an integer (whole number) \n";
  cin >> number1;
  cout<< "Please enter another integer (whole number)
\n";
  cin >> number2;
  // compute the answers
  product = number1 * number2;
  sum = number1 + number2;
  // print out the answers
  cout << "The sum of " << number1 <<"  and " <<
number2 << " is " << sum <<"\n";
  cout << "The product of " << number1 <<"  and " <<
```

```
number2 << " is " << product <<"\n";

  return 0;
}
```

Step 2: Compile the code by typing in *bcc32 example02-06.cpp* at the command line.

Step 3: Run the program by typing in *example02-06*. You should see an image much like the one displayed in Figure 2.6.

There are a number of items in this program that you should be aware of. The first is the use of the `cin` command. After declaring integer variables to hold the input, it is also necessary to use the command `cin >> number1`. That expression literally takes the input from the keyboard and places it into the variable `number1`. In this code you see the same thing done with the variable `number2`. Also notice that when you take in input you don't need to do anything to tell the compiler what kind of variable you want to put the data in. The `cin` command can see what kind of variable is to the right of the >> sign, and will convert the data to the proper type, if possible. Obviously if you try to put *a* into an `int`, you will have a problem.

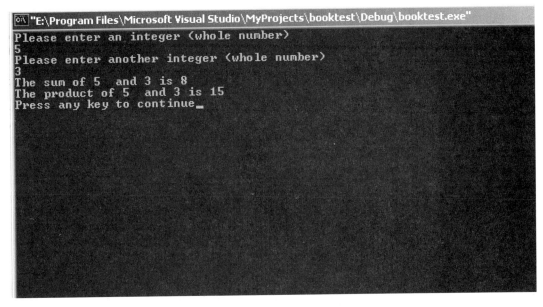

FIGURE 2.6   Input/Output and basic operators.

WATCHOUT!

Make sure that the variable you use for input matches the type of data you asked the user to enter. Do not ask for a decimal value then try and put that into an integer variable!

The next item that you should direct your attention to is the use of compound `cout` statements. You can output just about as much as you like with the `cout` command, in a single statement. All you must do is to separate the items you wish to output with << characters. Using this technique, you will be able to combine variables and text to form the output to the screen that displayed the answers. This is an important technique to remember. You will see it used many times in this book.

These simple examples demonstrate the basics of C++ input and output. With just these simple techniques you can handle input from the keyboard as well as output to the monitor screen.

Other Input Options

As you have already seen, the `cin` operator allows you to input data from the keyboard. However, it is not the only method or necessarily the best for all situations (but it is the simplest). The two other most commonly used methods are `get` and `getline`. The `get` function will retrieve a single character, no matter how much the user types in. The `getline` function will retrieve a certain number of characters, a number that you specify. This is particularly important when putting the data into an array.

HINT!

Arrays are covered in detail in Chapter 3. For now you should simply think of an array as a string of variables, as you saw with the character array, buffer.

Simply using the `cin` will allow the user to try and put more characters into the array than it can hold. This is referred to as *overflowing*. The get-line function allows you to specify how many bytes you will get from the user's input. Each character the user types takes up one byte. So if, in the `getline` function, you specify four bytes, and the user types in the word *computer*, you will only retrieve *comp*, the first four letters. In the following example you will see the `getline` function used with an array.

Example 2.7

Step 1: Write the following code into your favorite text editor.

```
#include <iostream>
using std::cin;
using std::cout;
int main()
{
 char text[10];
 cout << "Please enter a word\n";
 cin.getline(text,10);
 cout << text << endl;
 return 0;
}
```

Step 2: Compile and run the program.
Step 3: Type in more than 10 characters. You should see something like what is depicted in Figure 2.7.

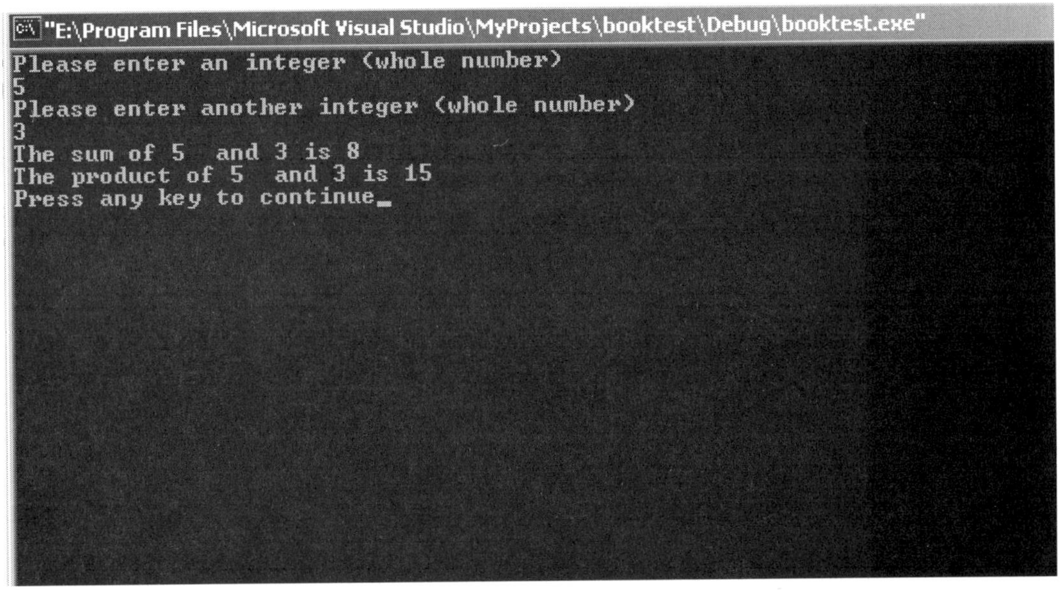

```
"E:\Program Files\Microsoft Visual Studio\MyProjects\booktest\Debug\booktest.exe"
Please enter an integer (whole number)
5
Please enter another integer (whole number)
3
The sum of 5  and 3 is 8
The product of 5  and 3 is 15
Press any key to continue_
```

FIGURE 2.7   The `getline` function.

You should notice that only 10 characters are placed in the character array, no matter how many you enter. In fact, you should probably prove this to yourself by deliberately typing in more text. This means that `get-line` is the safe choice if you want to enter a string of characters into an array.

IOSTREAM DETAILS

You achieved the previously demonstrated input and output functionality by importing a file called `iostream`. What exactly is in `iostream` and what does it do? The `iostream` file imports the `iostream` class. (Don't worry about what a class is just yet, you will learn all about this later. For now, think of a class as a bunch of related functions grouped together.) This class provides the basic capability for sequential and random-access I/O.

The `iostream` header file essentially uses streams, which are essentially a flow of characters. These characters may be 1 byte or 2 byte ("wide"), the latter are only necessary for languages with many characters.

USING THE STANDARD NAMESPACE

`Namespaces` allow C++ programmers to group a set of global objects or functions under a single name. The general format of a `namespaces` is the following.

```
namespace identifier
{
 namespace-body
}
```

The word *identifier* represents any valid identifier and `namespace-body` is the set of objects or functions that is included within the `namespace`. An example follows.

```
namespace mynamespace
{
 int a, b;
}
```

In the code segment shown here, *a* and *b* are normal variables integrated within the namespace named mynamespace. To access these variables from outside the namespace we have to use the scope operator ::. For example, to access the previous variables we would have to put:

```
mynamespace::a
mynamespace::b
```

The use of namespaces is particularly handy in cases where there is a possibility that a global object or function can have the same name as some other object or function. Using namespaces allows you to specify which one you are referring to. Let's look at a code segment that shows how namespaces can help with cases where two variables have the same name.

```
#include <iostream>
using std::cout;
using std::endl;
namespace nameone
{
  int myvar = 5;
}
namespace nametwo
{
  float myvar = 9.6f;
}
int main () {
cout << nameone::myvar << endl;
cout << nametwo::myvar << endl;

return 0;
}
```

HINT!
. . . .

After the 9.6, there is a small lowercase letter *f*. This is to denote that this number is a float. If you do not put that there, many C++ compilers will assume that it is a double.

Without the use of namespaces you would get an error trying to use two variables with the same name.

The most common use of namespaces, is by far with the standard namespace. When you include iostream you have been identifying each item you wish use. However, you don't have to do that if you simply specify that you are using the standard namespace. The following is code snippet that includes the standard namespace.

```
#include <iostream>
using namespace std;
```

You use namespaces by simply executing the following statement.

```
using namespace namespacename;
```

Now you can access the elements contained in that namespace simply by calling their names. It's a little like using the include statements we have already seen in this chapter.

```
// using namespace example
#include <iostream>
namespace nameone
{
int myvar = 5;
}
namespace nametwo
{
float myvar = 9.6f;
}
int main ()
{
using namespace nametwo;
cout << myvar << endl;
return 0;
}
```

In this case, because you stated explicitly that you were using namespace nametwo, whenever you reference the variable myvar, it will be the version contained in namespace nametwo.

You may be asking what this has to do with console input and output, and that's a good question. This topic is covered here because the standard namespace is what allows you to use names such as cout, cin , and endl. *Visual* C++ will allow you to use the .h extensions and not include the standard namespace. However this is not standard and not widely supported, so you will normally need to include the namespace. In fact, it's best to just use the ANSI/ISO standards wherever possible because these standards are supported by ALL C++ compilers. This book uses only ANSI/ISO standard C++, with a few minor exceptions, which are noted in the text. With all of that said, you should plan on using the ANSI/ISO standard way of including the `iostream` header file.

```
#include <iostream>
using namespace std ;
```

ADDITIONAL FORMATTING

There are other ways of formatting text, other than the escape characters. Many of these are new to C++ and are not supported in C. In addition to formatting, your C++ compiler can also convert values of various types into characters. You can control formatting with certain C++ commands such as the following:

- `cout << endl;`—Newline.
- `cout << flush;`—Flush buffer.
- `cout << hex;`—Base 16.
- `cout << dec;`—Base 10.
- `cout << setprecision(5);`—Sets floating point format.
- `fill(x)`—Pad fields with the x character. (You can use any character you wish.)
- `width(i)`—Sets the field width to the value represented by the integer *i*.
- `precision(i)`—Sets the number of significant digits displayed in floating point numbers.
- `setf(ios::flagname)`—Sets the various flags mentioned in the following section. You call the `setf` function and pass to it the word `ios::` followed by the actual flag you wish to set.

The following items are flags, rather than functions. A flag is a preset value that is used to determine a course of action. Flags are normally integer values, but often are given a name, or label, to make their value more readily apparent to human programmers. In this case, the flags determine how the `setf` function will format the output. This means that you simply set them to some value rather than call them, or pass them, parameters.

- Floatfield—A flag to set the style for floating point numbers: scientific (exponential notation) or fixed
- Scientific—A flag to set the style for scientific notation
- Adjustfield—A flag to set alignment in fields such as left or right aligned

This next example illustrates several of the various formatting functions and properties that we have just discussed. This should allow you to see these items in action and get a much better feel for them.

Example 2.8

Step 1: Type the following code into your favorite text editor.

```
#include <iostream>
#include <iomanip> // This header file is required for
    // the setprecision manipulator
using namespace std; // this is a shortcut. Instead of
// including each item in the std
// namespace, just include them
// all!
int main()
{
 cout << 2002 <<endl;
 cout << "In hex " << hex<< 2002 <<endl;
 cout.setf(ios::scientific,ios::floatfield);
 cout <<987.123456 <<endl;
 cout << setprecision(3) << 987.123456 <<endl;
 cout.fill('X');
 cout.width(10);
 cout << 1234 <<endl;
 cout.setf(ios::left,ios::adjustfield);
 cout.width(10);
 cout << 1234 <<endl;
 return 0;
    }
```

Step 2: Compile your code.

Step 3: Run the executable. You should see something similar to the image shown in Figure 2.8.

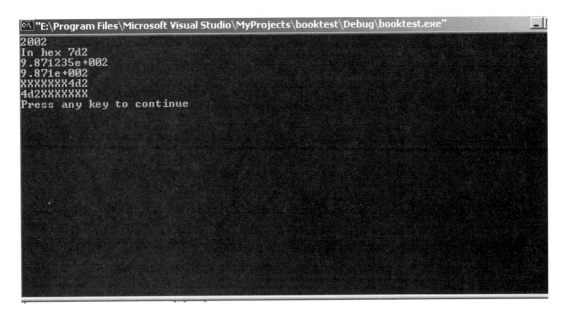

FIGURE 2.8   Formatting.

This section is meant to introduce you to these formatting techniques. You will see many of them used again throughout the book, so you will get a chance to get more comfortable with them as we go.

SUMMARY

This chapter is essential for you as a C++ programmer. In this chapter, you learned how to capture input from the keyboard and redirect that input into a variable. You also learned how to display that data to the screen. The use of the `cin` and `cout` functions is absolutely essential to any C++ program. Although simple and straightforward, the concepts of this chapter are of critical importance. Make absolutely certain that you have thoroughly understood them before you continue. This chapter also introduced you to some basic formatting commands and escape codes.

Another item you saw in this chapter was the use of `namespaces`. You may program for quite some time without needing to use more than the standard `namespace`. However, you will see nonstandard `namespaces` used in code samples you find either on the Internet or from other sources. So it does behoove you to become at least basically familiar with them.

REVIEW QUESTIONS

1. How do you cause the cursor to move to the next line when performing screen output?
2. What does the escape key \b do?
3. How would you direct input from the keyboard to a variable called `amountentered`?
4. To use `cin`, you must import what file?
5. Is this statement/expression correct?
   cout>> "Hello World";
6. What does the escape key \t do?
7. How would you output a literal text such as "The answer is" and the variable that holds the answer (let's call it answer) on the same line?
8. How would you create a new line if you were writing code in C instead of C++?
9. How would you put literal quotes in your output?
10. What is the purpose of the `setf` function?

3

Arrays, Strings, and Bitwise Operations

IN THIS CHAPTER
• • • • • • • • • • • •

- Arrays and Strings
- Character Arrays
- Using and Formatting Strings
- Bitwise Operations

In Chapter 1 you were introduced to variables. In Chapter 2 you used these variables in different ways, and the concept of an array was introduced. This chapter explores arrays in greater detail. This chapter also explores strings. Strings and arrays are closely related topics. Also included in this chapter is an introduction to bitwise operations.

ARRAYS AND STRINGS
• • • • • • • • • • • • • • •

Now you have seen the basic data types you can work with in C++. These will be adequate for most situations. However, they will not be the ideal solution for all situations. In some situations, you have closely related data that should logically be grouped under a single name. For example, if you wished to hold the temperature for 20 consecutive days, all the values could logically be called "temperature," but how do you have one variable hold 20 different values at the same time? The answer is an array. An array is basically a variable that can hold multiple values. A more formal definition would be an array is a series of variables of the same type, referred to by the same name, and contiguous in memory. Each variable in the array has the same name, and is differentiated by a number that indicates which element in the array it is. The following example declares an array.

```
float temperature[20];
```

What happens with this statement is that a variable named `tempera-ture` is created, but rather than just the space for a single float value being allocated, space is allocated for 20 different floats. They are loaded into memory right next to each other in order starting with zero as the first element: `temperature[0]`, `temperature [1]`, `temperature[2]`, and so on. Those floating point values can be accessed using a number to designate the element in the array you wish to address, like the following example.

```
temperature[0], temperature[1], temperature[2]….temperature[19]
```

Note that the temperature array stops at 19. It begins at 0 and has 20 elements, thus stopping at 19. All arrays begin with zero. If you define an array, the number you put in the brackets is the number of elements the array will have, beginning with zero.

Another way to define arrays is to say that they are simply a series of elements of the same type placed consecutively in memory. These elements can be individually referenced by adding an index to a unique name. In practical terms, this means that if we declare an integer array with five elements, then we can store five values of type `int` without having to declare five separate variables, each one with a different identifier. Instead, we can use an array to store the five different values of the same type.

You can access an element of an array the same way you would access any other variable; you simply have to provide the index that identifies which element of the array you wish to access.

```
temperature[5] = 89.5;
```

WATCHOUT!

If you attempt to access an element of the array that is beyond the bounds that you declared the array, most C++ compilers will let you try. . . . Unfortunately, this can lead to serious problems. The following example uses a value that is beyond the actual range of the array.

Example 3.1

Step 1: Enter the following code into your favorite text editor.

```
#include <iostream>
using namespace std;
int main()
{

  int myarray[5];
  myarray[10] = 99;
  cout << myarray[10] << endl;
  return 0;
}
```

Step 2: Compile and run the code. You should see something similar to what is depicted in Figure 3.1.

When you write values beyond where you defined the array, there is no way to know where that variable was stored. Basically it is stored at some spot in memory. Because we tried to access an element that was six elements beyond an array of integers, what happens is that the next space in memory that is 24 bytes (six elements multiplied by four bytes per integer) is overwritten with the number 99. This could be memory that contains data that is necessary for some other program. Writing beyond the limit of your array is called *overflowing the buffer*. This can lead to serious problems because you might be overwriting some other program's data. You should always be careful in your code to avoid attempting to access any element beyond the limits that you declared when you created your variable.

Before moving on, it is important that you fully understand what an array is. Many beginning programmers make this more difficult than it

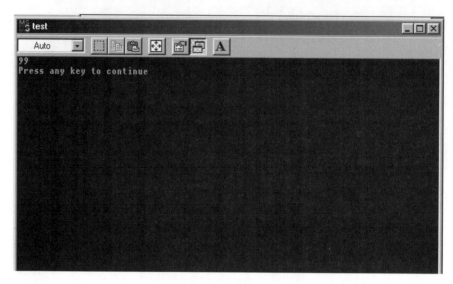

FIGURE 3.1 Overflowing arrays.

really is. An array is just another variable. However, instead of holding a single value, it holds a series of values. This is useful when you have data that is related. You can store each item in an element of the array. Therefore, an array is simply a series of variables with the same name and type, each designated by an index number, which identifies which element of the array we are addressing. You will notice that the past few paragraphs have given you a few different but similar definitions of an array. This is intentional. It is hoped that at least one of these definitions will help you to understand the concept. To illustrate how to access each element in an array, consider the following code fragment.

```
istudentsage[5];
studentsage [0]  =  14;
studentsage [1]  =  17;
studentsage [2]  =  15;
studentsage [3]  =  14;
studentsage [4]  =  16;
```

You can see that an array named studentage is created. It is five elements long, ranging from mystudent[0] to mystudent[5]. We then place a value in each of the elements of this array. You could, of course, have made five different int variables, each with a different name. However, because they all contain the same kind of information, an array makes better sense. Let's look at the following example.

Example 3.2

Step 1: Write the following code into your favorite text editor. Save it as *03_02.cpp*.

```
#include <iostream>
using namespace std;
int main()
{
    float temp[5];
    cout << "Please enter the temperature for day one \n";
    cin >> temp[0];
    cout << "Please enter the temperature for day one \n";
    cin >> temp[1];
    cout << "Please enter the temperature for day one \n";
    cin >> temp[2];
    cout << "Please enter the temperature for day one \n";
    cin >> temp[3];
    cout << "Please enter the temperature for day one \n";
    cin >> temp[4];
    cout << " The temperatures for the first five days is ";
    cout << temp[0] << ", " << temp[1] << ", "<< temp[2]
 << ", "<< temp[3] << ", ";
    cout << temp[4] << endl;
return 0;
    }
```

Step 2: Compile and run the code.

You should see an image, much like the one shown in Figure 3.2.

This example is pretty simple but it does show you the basics of declaring an array, filling an array, and printing an array.

It is also important to initialize arrays to some default value. A default value is simply some starting point value. It is commonplace to give all numeric variables a value of zero. When declaring an array that has *local scope* (local scope refers to anything that is declared within a function), if you don't specify otherwise, it will not be initialized. This means that its content is undetermined until we store some values in it.

If we declare a *global* array (a global variable is one which is declared outside of any individual function), its content will be initialized with all

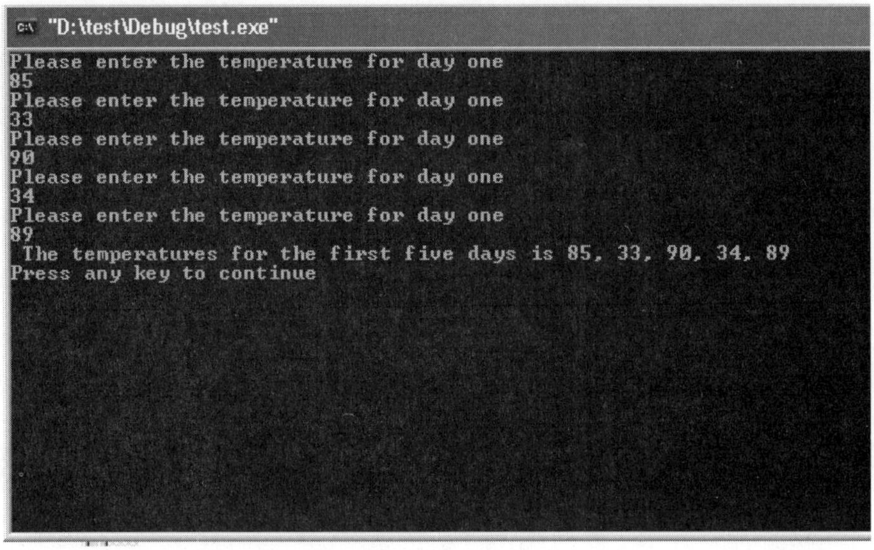

FIGURE 3.2   Using arrays.

its elements filled with zeros (if it's a numeric data type). What this means in practical terms is that arrays declared within a function must be initialized to some default value. Those declared outside of function don't have to be initialized, but may be if you wish. If you want to play it safe, initialize all arrays to some default value, regardless of where they are declared. This initialization can take place in several ways. The most common way is done when you first create the array.

```
int myarray [5] = { 10, 5, 10, 20, 15 };
```

The other way to initialize an array is to use a loop and loop through each element, giving it a value. Chapter 5 discusses loops in greater detail. For now you can simply use the first method for initializing arrays. It is, by far, the most common method.

The arrays you have seen are called one-dimensional arrays, because they only have one dimension. However, you can have more dimensions to your array. Let us examine the two-dimensional array. When you declare a two-dimensional array you must declare two different indexes, as you see in the following example.

```
int myarray[4][4];
```

This expression creates an array of integers that is four integers by four integers, thus holding a total of 16 integers. You access each element of a two-dimensional array by specifying two numbers. For example, the first element of this array would be myarray[0],[0]. The following table (Table 3.1) clarifies the positioning of elements in a two-dimensional array.

TABLE 3.1   Two Dimensional Array Elements

0,0	0,1	0,2	0,3
1,0	1,1	1,2	1,3
2,0	2,1	2,2	2,3
3,0	3,1	3,2	3,3

It is possible to create arrays with any number of dimensions that you see fit. However, it is rare to see arrays of more than two dimensions. There are several reasons for this. The first reason is that such multidimensional arrays quickly become very unwieldy for the programmer. Secondly, they quickly tax a system's resources. Consider that a single dimensional array of four integers has 4 elements and takes up 16 bytes (4 bytes per integer multiplied by 4 integers). A two-dimensional array of 4 integers by 4 integers has 16 elements and uses 64 bytes (16 bytes per row by 4 rows). But a three-dimensional array of $4 \times 4 \times 4$ integers would have 64 elements and would take up 256 bytes. In most cases you will encounter single arrays, and occasionally you will see two-dimensional arrays. It would be rare for you to need to work with arrays of higher dimensions. Perhaps it would be useful for you to see an actual two-dimensional array in action. This example will use a 4 by 7 float. This represents 4 weeks, at 7 days per week, and the temperatures recorded on each day. This single, two-dimensional array can store the temperatures measured for 28 consecutive days.

Example 3.3

Step 1: Enter the following code into your favorite text editor and save it as *03-03.cpp.*

```
#include <iostream>
using namespace std;
```

```cpp
int main()
{
    int monthlytemps[4][7];
    cout << "Enter the temp for week 1 day 1";
    cin >> monthlytemps[0][0];
    cout << "Enter the temp for week 1 day 2";
    cin >> monthlytemps[0][1];
    cout << "Enter the temp for week 1 day 3";
    cin >> monthlytemps[0][2];
    cout << "Enter the temp for week 1 day 4";
    cin >> monthlytemps[0][3];
    cout << "Enter the temp for week 1 day 5";
    cin >> monthlytemps[0][4];
    cout << "Enter the temp for week 1 day 6";
    cin >> monthlytemps[0][5];
    cout << "Enter the temp for week 1 day 7";
    cin >> monthlytemps[0][6];
    cout << "Enter the temp for week 2 day 1";
    cin >> monthlytemps[1][0];
    cout << "Enter the temp for week 2 day 2";
    cin >> monthlytemps[1][1];
    cout << "Enter the temp for week 2 day 3";
    cin >> monthlytemps[1][2];
    cout << "Enter the temp for week 2 day 4";
    cin >> monthlytemps[1][3];
    cout << "Enter the temp for week 2 day 5";
    cin >> monthlytemps[1][4];
    cout << "Enter the temp for week 2 day 6";
    cin >> monthlytemps[1][5];
    cout << "Enter the temp for week 2 day 7";
    cin >> monthlytemps[1][6];
    cout << "Enter the temp for week 3 day 1";
    cin >> monthlytemps[3][0];
    cout << "Enter the temp for week 3 day 2";
    cin >> monthlytemps[3][1];
    cout << "Enter the temp for week 3 day 3";
    cin >> monthlytemps[3][2];
    cout << "Enter the temp for week 3 day 4";
    cin >> monthlytemps[3][3];
    cout << "Enter the temp for week 3 day 5";
    cin >> monthlytemps[3][4];
    cout << "Enter the temp for week 3 day 6";
    cin >> monthlytemps[3][5];
    cout << "Enter the temp for week 3 day 7";
```

```
        cin >> monthlytemps[3][6];

        cout << "Enter the temp for week 4 day 1";
        cin >> monthlytemps[4][0];
        cout << "Enter the temp for week 4 day 2";
        cin >> monthlytemps[4][1];
        cout << "Enter the temp for week 4 day 3";
        cin >> monthlytemps[4][2];
        cout << "Enter the temp for week 4 day 4";
        cin >> monthlytemps[4][3];
        cout << "Enter the temp for week 4 day 5";
        cin >> monthlytemps[4][4];
        cout << "Enter the temp for week 4 day 6";
        cin >> monthlytemps[4][5];
        cout << "Enter the temp for week 4 day 7";
        cin >> monthlytemps[4][6];
        return 0;
}// end of main
```

Step 2: Compile and run the program.

You should notice several things about this little piece of code. The first thing you should notice is the way in which the array is being filled. This is actually a rather inelegant way to fill the array. When you encounter loops in Chapter 5, you will see more efficient ways to fill an array, and to access its elements. The important thing for you to notice now is that 28 temperature values are stored in a single array.

CHARACTER ARRAYS

One of the most common uses of an array is a character array. A character array is simply a string of characters that makes up one or more words. It is common to place such input into arrays of characters. Let's look at two examples.

Example 3.4

Step 1: Enter the following code into your favorite text editor.

```
#include <iostream>
using namespace std;
int main()
```

```
{
    char name[25];
    cout << "Please enter your name \n";
    cin >> name;
    cout << "Hello " << name << endl;
    return 0;
    }
```

Step 2: Compile and run the program. You should see something similar to what is shown in Figure 3.3.

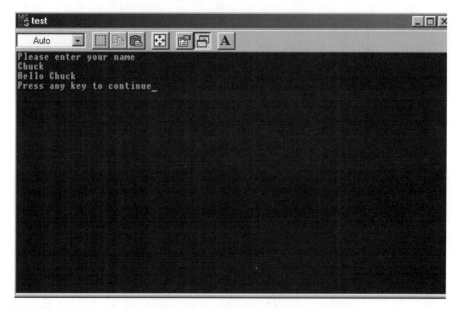

FIGURE 3.3 Character arrays.

This code is a rather straightforward and practical illustration of how to use arrays, specifically arrays of characters. There is only one problem. We declared the array to hold 25 characters. What happens if the user enters 30 characters? The answer is that your program may crash. However, there is a solution. In Chapter 2 you were introduced to the getline function and you were told that it was the safest way to input data into an array. Recall that the getline function specified what array to put the address in, and how many bytes to put in. If the user enters more data than that, then the extra characters are simply ignored.

Example 3.5

Step 1: Enter the following code into your favorite text editor.

```
#include <iostream>
using namespace std;
int main()
{

 char name[25];
 cout << "Please enter your name \n";
 cin.getline(name,25);
 cout << "Hello " << name << endl;
 return 0;
}
```

Step 2: Compile and run the code. You should see something similar to Figure 3.4.

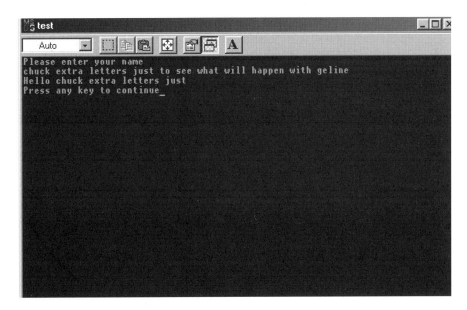

FIGURE 3.4   Using getline with character arrays.

USING AND FORMATTING STRINGS
• •

You have already seen several different variable types and we have used them for input and output. There is another type of variable we have not used yet. This is the string variable. A string is essentially a bunch of characters all together that you can use to store text. Up to this point we simply used an array of characters. You can still do this, but now you have other options. Now, C++ has a string variable that you can use. To use the string variable you will have to include the string header file. Here is an example.

Example 3.6

Step 1: Enter the following code into your favorite text editor.

```
#include <iostream>
#include <string>
using namespace std ;
using std::cout;
using std::cin;
int main()
{
 string s = "C++ is sooo coool! \n";
 cout <<s;
 return 0;
}
```

Step 2: Compile and run the program. You will see something similar to what is shown in Figure 3.5.

When you run this you will see how the string variable works. Essentially it is an open-ended character array you can use to store string data.

With C++, your string functions are defined in the string header file. Each string is an object that has methods and properties that you can use. Objects and classes will be covered in Chapter 10. For now, remember that an *object* is a special type of variable that has functions associated with it. These functions are called methods. The length property is a commonly used property of strings, which is easy to implement and understand.

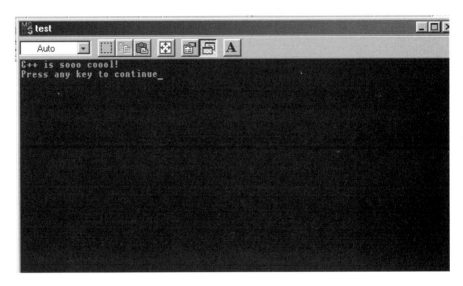

FIGURE 3.5   The C++ string.

Example 3.7

Step 1:

```
#include <iostream>
#include <string>  // include for C++ standard string
// class
using namespace std;
int main()
{
    string stringA = "C++";
string stringB = "Is Cool";
cout << "Length of stringA = " << stringA.length() <<
  endl;
cout << "Length of stringB = " << stringB.length() <<
    endl;
 return 0;
}
```

Step 2: When you run this code you should see something similar to
what is shown in Figure 3.6.

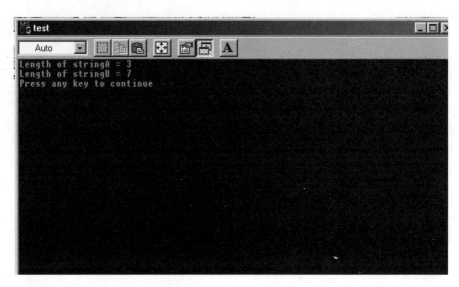

FIGURE 3.6   String operations.

You can see that, in C++, the string is an object with properties and methods. There are a number of properties and methods with this object; the most commonly used are summarized in Table 3.2.

TABLE 3.2   Properties and Methods of the String Class

Property/Method	Purpose
Length	Property that returns the length of the string.
Copy	This copies from one string to another.
Erase	This method erases the string.
Insert	This inserts a set of characters into a string.

Most C++ compilers still support C header files, and C functions. These C-style string functions are still used frequently. Therefore, a summary of some of these functions and an example are included here. The more important and commonly used functions from string.h are summarized in Table 3.3.

TABLE 3.3   String Functions

Function	Purpose	Format
strcpy	To copy one string to another.	char * strcpy (char * dest, const char * src);
		Note: The * is a special operator that will be explained in detail in the chapter on pointers.
strlen	To return the length of a string.	size_t strlen (const char * string);
strncat	Append a sub string onto another string.	char * strncat (char * dest, const char * src, sizet_t num);
strstr	Find one string inside of another string.	char * strstr (const char * string1, const char * string2);

There are several other functions in the string header file, but these are the most commonly used functions. The following example illustrates the use of these functions.

Example 3.8

Step 1:

```
#include <iostream>
#include <string >
using std::cout;
using std::cin;

int main()
{
 char firststring[40],secondstring[40],thirdstring[40];
 int size;
 cout << "Please enter a word \n";
 cin.getline(firststring,40);
 cout << "Please enter another word  \n";
```

```
cin.getline(secondstring,40);
size = strlen(firststring);

 strcat(firststring,secondstring);
 cout << "The length of the first string you entered
is" << size << "\n";
 cout << "Both strings you entered are " <<
 thirdstring<< "\n";
 return 0;
}
```

This example, although simple, should show you how to use some of the functions you will find in the string header file.

HINT!

The previous example is using C-styled string manipulations. This is actually supported by many of the commonly used C++ compilers.

BITWISE OPERATIONS

The bitwise operators are a bit more difficult to understand than standard operators (pun intended). The first order of business is to define what a bit is. The computer ultimately stores everything as either a 1 or a 0. A single value of either 1 or 0 is called a bit. Essentially C++ allows you to work directly with the binary equivalents of numbers, to work with bits. This is useful in many computer applications, especially in the telecommunications industry. You will also see these operations in a later chapter when we explore basic encryption algorithms. Explaining bitwise operations necessitates first explaining binary numbers.

The numbers you are familiar with in day to day usage are decimal numbers, base 10 numbers. The base 10 number system is only one possible number system. You can use base 2, base 5, base 16, or any other base you wish. The reason most human societies have always used base 10 number systems is that we have ten fingers. Therefore, our ancestors had an easier time counting in tens, thus number systems started out being based on 10. A computer, however, "speaks" in on or off, base 2 numbers (also called binary numbers). For this reason a basic understanding of binary numbers is essential for programming. In the base 2 number system you can only have 0's and 1's. (Notice you don't actually have any 2's, just as the base 10 system only has 0's to 9's.) What this means is that after you have placed a zero or a one in a particular place, to increase the number you have to move over to a new place. This table (Table 3.4) might help clarify that a bit.

TABLE 3.4   Binary Numbers

Base 10 Number	Base 2 Equivalent
0	0
1	1
2	10 (Note this is a 1 in the 2's place and a 0 in the 1's place, or one 2 and no 1's.)
3	11 (This is a 1 in the 2's place and a 1 in the 1's place, thus equaling one 2 and one 1, which is 3.)
4	100 (This is a 1 in the 4's place, a 0 in the 2's place, and a 0 in the 1's place.)
5	101 (This is a 1 in the 4's place, a 0 in the 2's place, and a 1 in the 1's place.)
6	110 (This is a 1 in the 4's place, a 1 in the 2's place, and a 0 in the 1's place.)
7	111 (This is a 1 in the 4's place, a 1 in the 2's place, and a 1 in the 1's place. And 4 plus 2 plus 1 equals 7.)
8	1000 (This is a 1 in the 8's place with a 0 in the 4's, 2's, and 1's places.)

Hopefully you are noticing a pattern here. The *places* in any number system are essentially the number system's base raised to some power. For example, in the base 10 system you have a 1's place, 10's place, 100's place, 1000's place, and so on. If you think about this for just a moment you should realize that 1 is simply 10^0, 10 is 10^1, 100 is 10^2, 1000 is 10^3, and so on. The same is true for the base 2 number system, where you have a 1's place, 2's place, 4's place, 8's place, 16's place, and so on. 1 is simply 2^0, 2 is 2^1, 4 is 2^2, 8 is 2^3, 16 is 2^4, 32 is 2^5 and so forth.

In computers everything is stored in groups of eight bits, called bytes. For this reason all binary numbers are actually represented by an eight-digit series of bits. So that one is represented as 00000001, two as 00000010, three as 00000011, and so on.

Now that you have a basic understanding of what binary numbers are, we can explore bitwise operators. Bitwise operators work on integers as if they where binary numbers. Let's examine the bitwise shift operators first. A bitwise left shift literally shifts all the bits to the left the specified number of places. Consider the following shift operation.

```
4 << 1
```

This means to take the binary equivalent of 4 (which is 00001000) and shift all the bits over 1 space. This makes the number 00010000. Notice that this shifts it over to an entirely new place, the 8's place. So operation 4 <<1 actually yields 8. The right shift operator (>>) simply shifts the numbers to the right rather than the left.

The other two bitwise operators (& and |) are of particular use in a variety of applications including such diverse areas as *Windows* programming, telecommunication, and encryption, as you will see later in this book. The & operators simply asks whether the digits in both numbers are a 1 (and not a 0). For example, 4 & 2 would actually be evaluated in binary as 00001000 & 00000010; to see how this works, compare them vertically.

```
00001000
00000010
```

You can now see that this would yield 00000000, because in neither case do both numbers have a 1 in the same place. To further clarify this point, let's compare two other numbers, 5 & 3. In binary this would be the following.

```
00001001
00000011
```

This would yield 00000001 because only in the final place do the two numbers have a 1 in the same place.

The | (or) operator asks whether either number has a 1 in that particular place. So if you compare 5 | 3 you would have the following.

```
00001001
00000011
```

And that would yield 00001011 (which is 7 in decimal numbers).

The ^ operator is called the exclusive or, or XOR operator. It asks if there is a 1 in one of the numbers but not both, at a given place. For example if you consider 7 and 4 and you XOR them you see the following.

```
00000111
00000100
```

You get 00000011 or 3 because the XOR operation asks if there is a 1 in the place of one number but NOT the other. In other words, the XOR operation is exclusive to one or the other number, but not both, thus the name. The following example will allow you to AND, OR, and XOR integers.

Example 3.9

Step 1: Enter the following code into your favorite text editor.

```
#include <iostream>
using namespace std;
int main()
{

  int num1, num2,iand,ior,ixor;
  cout << "Enter an integer \n";
  cin >> num1;
  cout << "Enter another integer \n";
  cin >> num2;
  iand = num1 & num2;
  ior = num1 | num2;
  ixor = num1 ^ num2;

  cout << num1 << " AND " << num2 << " is " << iand <<
  endl;
  cout << num1 << " OR " << num2 << " is " << ior <<
  endl;
  cout << num1 << " XOR " << num2 << " is " << ixor <<
  endl;
  return 0;
}
```

Step 2: Compile and run the code. You should see an image much like the one shown in Figure 3.7.

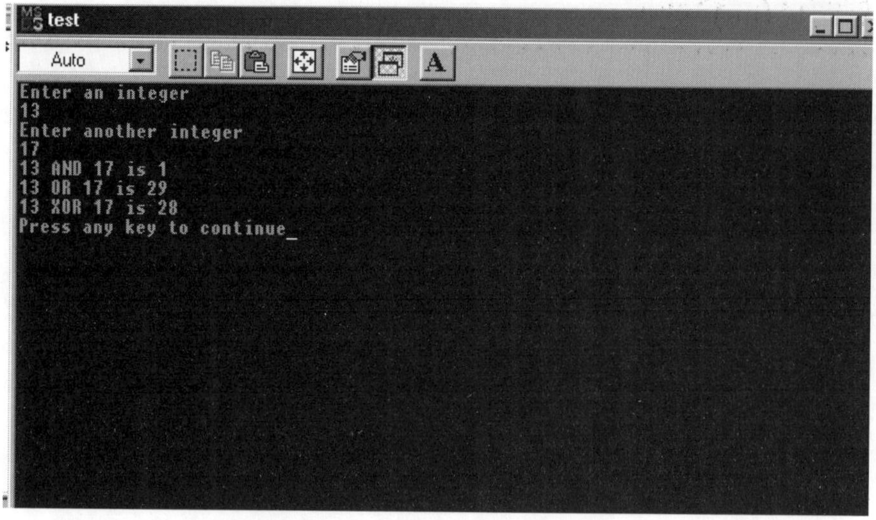

FIGURE 3.7 Bitwise operations.

These binary operators are going to be of particular importance to you later when you encounter encryption, but they are also used in other types of programming. It is also important to understand binary numbers for the simple reason that this is what the computer ultimately translates everything you enter into. The computer only understands binary numbers, so anything you enter into it gets translated into binary.

SUMMARY

This chapter introduced you to strings and arrays. These are vital programming topics that you should be familiar with. Arrays are commonly used in all programming languages. This chapter also introduced you to the concepts of binary numbers and bitwise operations. These types of operations are quite common in C and C++ programming.

REVIEW QUESTIONS

1. What is an array?
2. What does the ^ operator do?
3. What is a multidimensional array?
4. A string is actually an _____ of _____.
5. List four string methods.
6. How would you initialize this array to all 0's: int myarray[4];
7. All arrays start at what index number?
8. What is the purpose of the cstring header file?

Functions

IN THIS CHAPTER

- Basic Structure
- Creating and Calling Functions
- Function Overloading
- Header Files
- Passing Values by Reference
- Built-in Functions
- Math Operations
- Time
- Random Numbers

Chapter 1 provided a brief introduction to the concept of functions. A main function and the basic structure of a function were shown. This chapter will go into functions with much greater detail. You will learn how to create and use a variety of functions. In this chapter you will also be introduced to advanced concepts such as function overloading.

BASIC STRUCTURE

The first item to address is to explain exactly what a function is. A *function* is essentially a block of code that can be called by name, may be passed parameters, and may return a value. Another way to put that would be to say that a function is a group of related statements/expressions that work together to accomplish some goal and are grouped under a common name. Let's take a look at the main function that all C and C++ programs must have. By dissecting it, you may get a better idea of what a function is and how one is built.

```
int main()
{
   return 0;
}
```

The first line is called the function declaration line. It contains three major parts. Those parts are the return type, name, and parameter list. The return type simply tells you what kind of data this function will return. One way to think of it is to realize that you create functions to perform some goal. Whatever answer is generated from that function is what is returned. If you create a function to compute the area of a circle, then the return value would be the actual area computed. The return type for most functions can be any valid data type. It can be an int, float, bool, long, and so on. However, the main function must return an int. If the function does not return anything at all, then its return type is "void."

HINT!

ANSI/ISO standards dictate that main must return an int. You will see some programmers return a void, but this is nonstandard. In this book, all main functions will return an int value. A return of 0 indicates that all is OK; a return of one indicates that something went wrong.

The next portion of the function declaration is the name. The primary function, the function that starts the program must be named main. Other functions can have any name you wish, provided the name follows the same conventions that variable names do. It is, however, a good idea to give the functions a name that indicates what they do. For example, if your function squares a number, give it a name such as the names you see here.

- square_num
- number_squared
- num_2

The last part of the declaration line is the parameter list. These are things you give the function. If, for example, your function squares a number, then you have to give it the number you want squared—that number is a parameter. You can have no parameters, one parameter, or several parameters. Also recall that another name for a parameter is an argument.

All of that makes up the declaration line of the function. What is left now is relatively simple. You have brackets { } enclosing the function, the statements/expressions that you wish the function to execute, and, finally, you return a value. Remember that brackets are used to denote boundaries around a block of code. Right now that block of code is a function. In later chapters you will see other blocks of code (such as loops and switch statements).

WATCHOUT!

Remember that when returning a value, the value returned must, obviously, be of the same type that you declared as the function's return type. If you declared the return type as int, you cannot use return 1.22, or return true, as your return statements. You will get a type mismatch error.

CREATING AND CALLING FUNCTIONS

Obviously you could simply put everything into the main function and it would, indeed, compile and run, but this would quickly become unwieldy. In fact, your program would certainly become completely unmanageable and unreadable by the time it even reached a few dozen lines of code. Fortunately you don't have to do that. You can create separate functions to handle these other items. There are a few idiosyncrasies regarding function creation in C++. The following list summarizes the requirements.

1. The function must have a return type. It may be void if the function returns nothing.
2. The function must have a valid name.
3. The function must have parentheses for a list of parameters past to it, even if the parentheses are empty.
4. The function must either be written before it's called, or it must be prototyped.

The first three you already know about, but the fourth rule may sound quite confusing. What does it mean to prototype a function? Essentially the problem is that the C++ compiler cannot call a function it does not know about. You have to let the compiler know what that function is and what it does before you can call it. One way to do this is to make sure you write out the function in your file, before you call it. This quickly becomes quite cumbersome. Therefore, C++ provides another way to do this. That method is prototyping. All you do to prototype a function is to write its declaration line at the beginning of the source file. This would probably be much clearer with an example.

Example 4.1

In this example we will create a function that simply cubes any number passed to it and returns the answer. That function is prototyped and then called from the main function.

Step 1: Open your favorite text editor and type in the following code. You will then save it to a file called *04-01.cpp*.

```cpp
// Include statements
#include <iostream>
 using namespace std;
// function prototypes.
float cube_number(float num);
int main()
{
 float number;
 float number4;
 cout << "Please enter a number \n";
 cin >> number;

 number4 = cube_number(number);
 cout << number << " cubed is " << number4;
 return 0;
}
float cube_number(float num)
{
 // this function simply takes a number, cubes it,
 // then returns the answer
 float answer;
 answer = num * num * num;
```

```
    return answer;
}
```

Step 2: Compile the code. (See Appendix D if you need more instruction on how to compile your programs.)

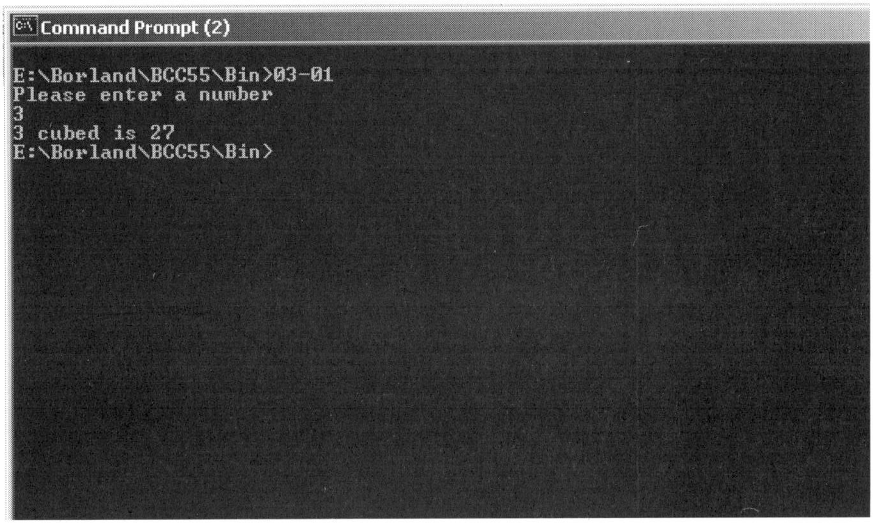

FIGURE 4.1   Basic functions.

Step 3: You can now run your program by typing in *04-01* at the command prompt. If all goes well (no compile errors), then you will see something like what is shown in Figure 4.1.

This code contains the essentials of writing, prototyping, and using functions. Let's examine it piece by piece to make sure you understand what is happening. The first line is simply an include statement. We include the `iostream` header file, so that you can do input and output (as you first saw in Chapter 2). The second line identifies that we wish to use the `standard namespace`, thus giving us access to `cout`, `cin`, `endl`, and any other function that may reside in `iostream`. Next is our prototype for our function. It is literally an exact copy of that functions declaration line. It lets the compiler know that there is a function later in the code. It's defined so that it takes certain parameters and returns a certain value. (If you see a call to this function, before you get to the part where the function is written, don't panic!).

HINT!

The main function does not need to be prototyped; its declaration line is built into C++. Also, functions that are included in one of the C++ header files do not need to be prototyped (such as cin and strcpy from Chapter 2).

Once you get into the main function of this program, you see a few variables declared. One variable, number, stores whatever number the user types in. The second variable, number4, is designed to store the answer that the cube_number function returns. When you call cube_number, because it returns a value, you must set the function on the righthand side of an equation with an appropriate variable. The following is a generic example. Some variable, of the same data type as the function's return type, is set equal to the function name. For example, float:

```
f = myfunc()
```

This example uses a variable of type double (the same type as the function is declared to return) and sets that variable equal to that function. This expression will cause the function to be called and its value stored in the variable in question.

You should also pay particular attention to the function cube_number. Notice that it does have a parameter. Beginning programmers sometimes have trouble understanding what to pass as a parameter. The simple rule of thumb is this: If you were going to ask some person to perform the same task you are asking the function to perform, what would you have to give them? The answer to that is the parameters you will need to pass to the function in question. If you were asked to cube a number, someone would have to give you the number to cube. Thus, if you ask a function to cube a number, you have to pass it that number. Parameters are simply values stored in variables that you pass to functions. They are the raw "stuff" that functions use to produce the product desired. Parameters are also frequently called "arguments." The words *argument* and *parameter* are used interchangeably.

Let's take a look at another example.

Example 4.2

In this example, the function takes two parameters to compute the area of a triangle. The formula for this is area = $\frac{1}{2}$ of the base of the triangle, multiplied by the height of the triangle.

```
A = 1/2 (b * h)
```

Step 1: Enter the following code in your favorite text editor. Save it in a file called *04-02.cpp.*

```cpp
#include <iostream>
 using namespace std;
// function prototypes
float triangle_area(float b, float h);
int main()
{
 float base, height, area;
 // get the user input
 cout << "Please enter the base of the triangle
\n";
 cin>> base;
 cout << "Please enter the height of the
triangle\n";
 cin >> height;
 area = triangle_area(base , height);
 cout << "That triangles area is " << area;
 return 0;
}
float triangle_area(float b, float h)
{
   float a;
   a = .5 * (b * h);
   return a;
}
```

Step 2: Compile the code.

Step 3: Run the code. You should see something similar to what is presented in Figure 4.2.

This example is a lot like the first example, with the exception that the function triangle_area takes two parameters. A function may take zero parameters, one parameter, two parameters, and so on. Theoretically you can have a long list of parameters (also called arguments), dozens if you wish. However, as a rule of thumb, if you need more than three or four parameters, you may wish to rethink your code.

You should note that with the exception of the two parameters instead of one, this program is very much like the first one. It begins with include statements, prototypes any functions other than main, has the main function, and then has other functions. You don't have to declare the main function first. However, it seems logical to put that function first, because

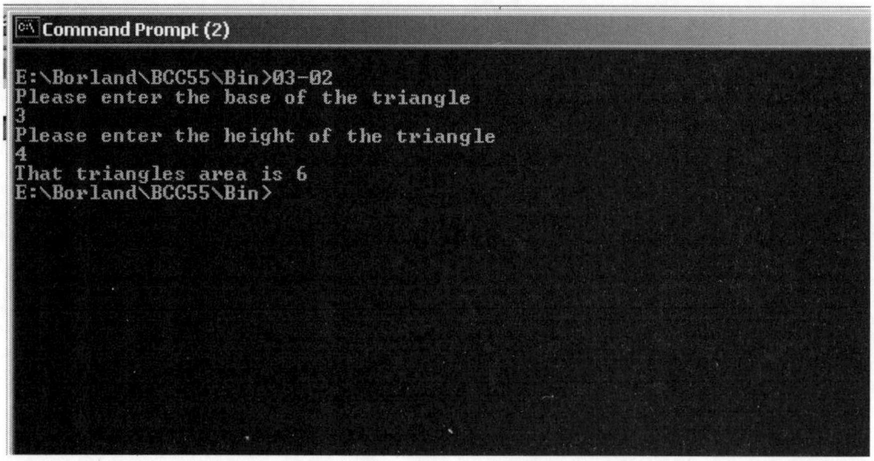

FIGURE 4.2   Basic functions 2.

that is where the entire program is going to start. Also remember that you can have a function call another function. In fact, in both of our examples, the main function called the second function.

FUNCTION OVERLOADING

C++ provides you with a powerful tool—overloading. You can overload a function so that the same name is shared by more than one function, but each function performs a different action. Function overloading is essentially when you have more than one function with the same name, but each performs different arguments. The first question that comes to your mind is probably: Why would you want to overload a function? There are several cases in which you may want to do this. For example, if you wanted to do an operation on some input, but the input came in different forms. The input might be an integer value or a floating point decimal value. You could overload the function in question, and have one version take an integer argument and the other a float argument. Let's take a second look at our number cubed example to illustrate this.

Example 4.3

Step 1: Open your favorite text editor and type in the following code. Save the code as *04-03.cpp*.

```
// Include statements
#include <iostream>
 using namespace std;
// function prototypes.
float cube_number(float num);
int cube_number(int num);
int main()
{
 float number;
 float number4;

 cout << "Please enter a number \n";
 cin >> number;
 number4 = cube_number(number);
 cout << number << " cubed is " << number4;
 return 0;
}
int cube_number(int num)
{
 int answer;
 answer = num * num * num;
 return answer;
}
float cube_number(float num)
{
 float answer;
 answer = num * num * num;
 return answer;
}
```

Step 2: Compile your code.

Step 3: Execute the code. You should see something like the image shown in Figure 4.3.

Notice that there are two functions named cube_number. They both take a single argument and return that argument cubed. However, the first function takes an integer and returns an integer answer, whereas the second function takes a floating point number and returns a floating point answer. The big question is how does the compiler know which function you are calling? The answer is simple, it knows which function you are calling based on what kind of argument you pass it. If you pass an argument with a decimal point, then it will call the function that takes a floating decimal point. If you pass an argument without a decimal point, then it will call the function that takes an integer.

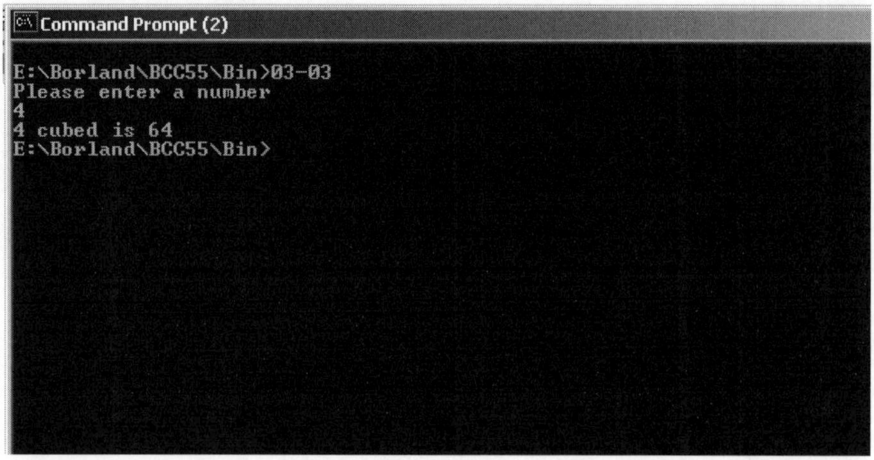

FIGURE 4.3   Function overloading.

This is why all overloaded functions MUST have different parameter types and/or a different number of parameters. You will see examples with different numbers of parameters later in this book.

HEADER FILES

So far we have only used header files that were built into C++. We have not used any files that we created. This section will change all of that. Header files are used to contain the prototypes for a function. (Other things can be placed in them, as we will see later in this book.). All you need to do is open your favorite text editor and place the prototype lines in that editor, then save the file with an *.h* extension.

Example 4.4

This example will show you how to create and how to use header files for function prototypes.

Step 1: Open your favorite text editor and type in the following code.

```
// function prototypes.
float cube_number(float num);
int cube_number(int num);
```

Save this file as *test.h*.

Step 2: Open a new instance of your favorite text editor and type the following code.

```
// Include statements
  #include <test.h>
#include <iostream>
 using namespace std;
int main()
{
 float number;
 float number4;
 cout << "Please enter a number \n:'
 cin >> number;
 number4 = cube_number(number);
 cout << number << " cubed is " << number4;
 return 1;
}
int cube_number(int num)
{
 int answer;
 answer = num * num * num;
 return answer;
}
float cube_number(float num)
{
 float answer;
 answer = num * num * num;
 return answer;
}
```

Step 3: Compile your code.

Step 4: When you run your code, you should see something like Figure 4.4.

This is a simplified example, but it does illustrate the basic points. Usually the actual functions would also be in a different file, if you were using a header file. However, in this case the purpose was to introduce you to creating your own header files.

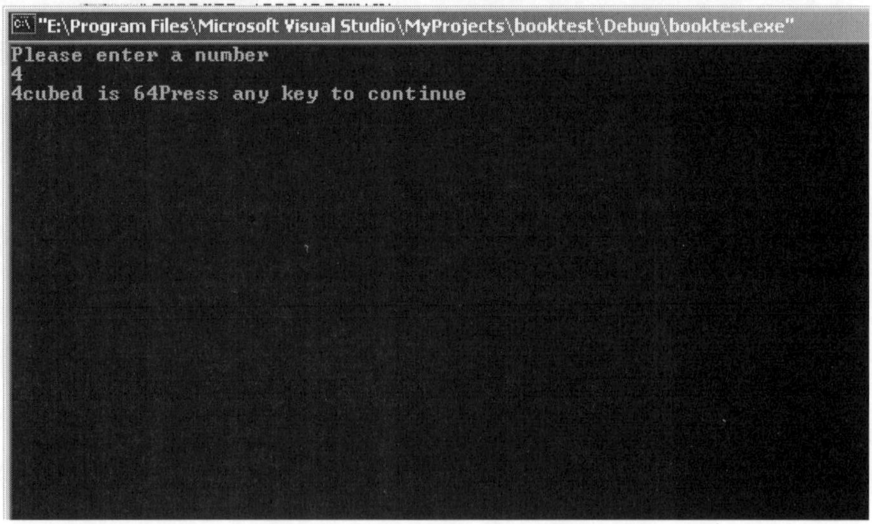

FIGURE 4.4   Header files.

PASSING VALUES BY REFERENCE

When you pass a value to a function you are really copying the contents of one variable to another variable—the other variable is the parameter of a function. This means the parameter is, essentially, a copy of the variable that was passed to the function. This means that you are not changing the value of the variable that is outside the function. That might sound confusing, so let's look at an example that illustrates this concept.

Example 4.5

Step 1: Place the following code into your favorite text editor.

```cpp
#include <iostream>
 using namespace std;
   void demo(float);
   int main ()
   {
float num1;
cout << "Please enter a number. \n";
cin >> num1;
cout << "Before the demo function your number is "
```

```
      << num1 << "\n";
   demo(num1);
   cout << "After the demo function your number is
      still" << num1 << "\n";
    return 0;
      }
   void demo(float number)
    {
    number = number * 4;
    cout << "Inside the demo function the number is now
   " << number << "\n";
    }
```

Step 2: Compile this code.
Step 3: Run the code. You should see something similar to the image shown in Figure 4.5.

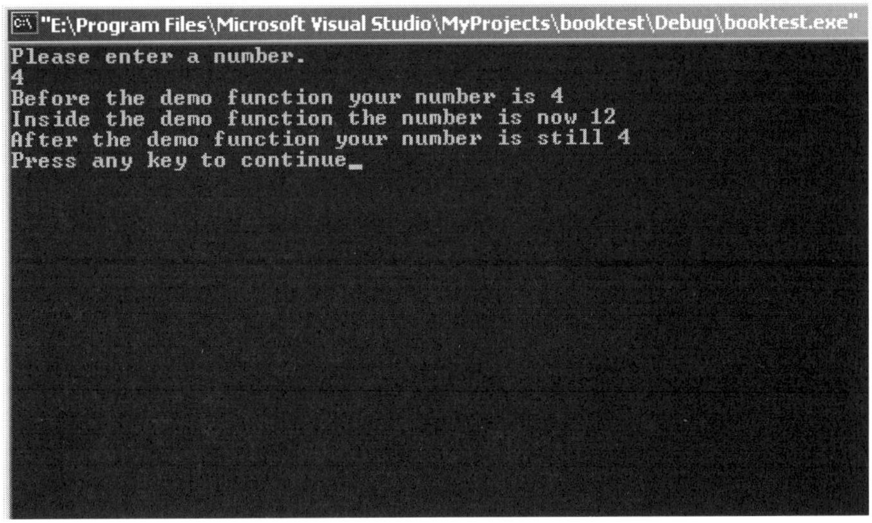

FIGURE 4.5 Passing variables.

As you can see from this example, what occurs inside the function has no bearing on the values of the variables outside the program. When you call the demo function, you take the value that is in num1 and place that value in the function parameter number. Number is an entirely new variable occupying its own space in memory and is separate from num1. This is called passing a variable by value. It is the standard way that variables are passed in C++ (as well as in *Java* and several other languages).

There is another way to pass a variable. That is by reference. When you pass by reference you don't simply copy the value of some variable to a parameter. What you pass is a reference to the variable's address in memory. What that means is that the parameter of the function is NOT a new variable. It is simply a new name for the variable. Both the variable name and the parameter name refer to the same place in memory, they refer to the same value. You accomplish this by simply adding the & or `address of` operator to your function parameter. The & means `address of` and it refers to the address in memory. Let's look at the previous example redone to pass-by reference.

Example 4.6

Step 1: Write the following code in your favorite text editor.

```cpp
#include <iostream>
 using namespace std;
void demo(float &number);
int main ()
{
  float num1;
  cout << "Please enter a number. \n";
  cin >> num1;
  cout << "Before the demo function your number is "
<< num1 << "\n";
  demo(num1);
  cout << "After the demo function your number is"
    << num1 << "\n";
return 0;
}
void demo(float &number)
{
 number = number * 4;
 cout << "Inside the demo function the number is now
    <DS>" << number << "\n";
}
```

Step 2: Compile the code.

Step 3: Run the code. You should see something like what is shown in Figure 4.6.

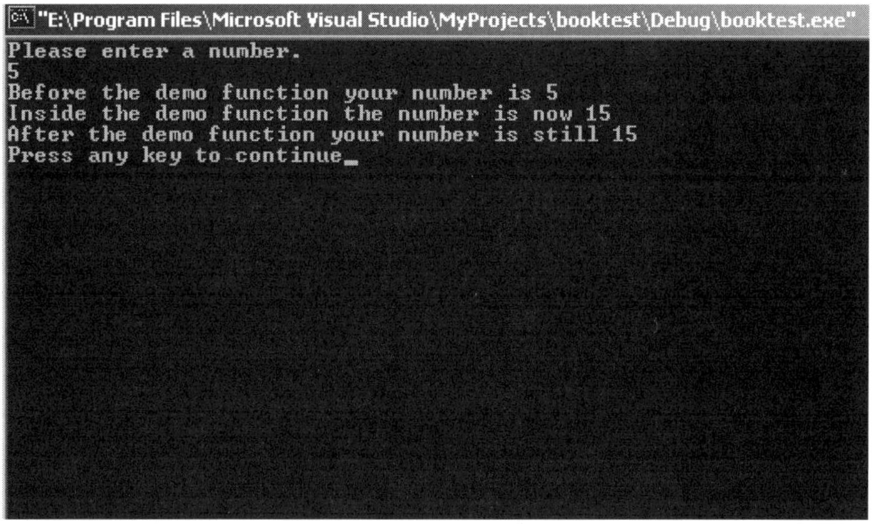

FIGURE 4.6   Passing with the & operator.

Notice that this time when the value is changed inside the function, it is also changed in the variable outside the function. This is because you passed the address of that variable to the function. The parameter name was just an alias for the variable. Whatever you did inside the function was actually happening to the variable.

BUILT-IN FUNCTIONS

In addition to the various functions you can create, C++ includes some useful functions you can use. Actually the term *built-in* is not exactly accurate, but it is close enough. These are functions that can be included in your program and then use. In fact, there are quite a few such functions. This book will show you some of the most commonly used functions and also show you some examples of how to use these functions. A few of the more commonly used functions are summarized in Table 4.1.

TABLE 4.1 C++ Functions

Function	Purpose
void *memcpy(*destination, *source, size);	This function copies whatever is in the source (often an array or structure [see Chapter 8]) to the destination. The size tells you how much to copy. To use this function you must either include the memory or string header files.
void * memset(*destination, replacement, size)	This function sets the destination to the character shown in replacement. This is often used to set an array to all zeros. To use this function you must either include the memory or string header files.
char* itoa(int value, char* buffer, int radix) Note: This one is NOT ANSI Standard C++ but is quite commonly used and supported by most C++ compilers.	This takes an integer and returns the character equivalent. The radix is the base of the integer you are converting it from (base 2, base 10, etc.).
int atoi(const char *string);	This function takes a character and returns the integer equivalent.
int tolower(int c);	Converts the character to lowercase. You must include the string header file for this function.
int toupper(int c);	Converts the character to uppercase. You must include the string header file for this function.

HINT!

memcpy and memset are frequently used with structures, which will be described in detail in Chapter 8. They can, however, also be used with arrays, which were introduced in Chapter 3.

The memcpy function shown in Table 4.1 is used to copy complex data types, such as arrays and structures, from one location in memory to another. You cannot simply copy one array to another as you would copy a standard variable. The following is an example.

```
int x,y;
x = 4
y = x
```

This works fine, and the contents of x are copied to y. However, this does not work with arrays.

```
int x[10] = {2,4,1,6,6,9,0,11,1,2};
int y[10];
y = x;
```

This will not work. It will generate a compiler error. So how do you copy one array to another? That is where memcpy comes in. To use memcpy, you must first include the memory header file.

```
#include <iostream>
#include <memory>
using namespace std;
int main()
{
    int x[10] = {1,2,3,4,5,6,7,8,9,0};
    int y[10];
    memcpy(y,x,10);
 return 0;
}
```

The function copies items from the second parameter to the first parameter, from the source to the destination. It copies the number of bytes you indicate in the third parameter, the size parameter. You will find these functions, and others, used throughout this text, and in many other C++ books. Let's take a look at an example that uses a few of these.

Example 4.7

Step 1: Type the following code into your favorite text editor.

```
#include <iostream>
 using namespace std;
int main()
{
 char a,u,l;
 int x;
 cout << "Please enter a number \n";
 cin >> a;
 x = atoi(&a);
 cout << "The character " << a << " is converted to
 the integer " << x << "\n";
```

```
cout << "Please enter a character \n";
 cin >> a;
 u = toupper(a);
 l = tolower(a);

 cout << a << " in upper case is " << u << " in
lower case is " << l;
 cout << "\n Press a key to continue ";
 return 0;
}
```

Step 2: Compile the code.

Step 3: Run the code. You should see something similar to what is shown in Figure 4.7.

```
"E:\Program Files\Microsoft Visual Studio\MyProjects\booktest\Debug\booktest.exe"
Please enter a number
5
The character 5 is converted to the integer 5
Please enter a character
a
a in upper case is A in lower case is a
 Press a key to continue Press any key to continue_
```

FIGURE 4.7   Assorted functions.

HINT!
••

The & operator is used with the `itoa` function. This is the `addressof` operator and means that we are passing the address of the variable rather than the actual value contained in the variable.

The previous example illustrated the most commonly encountered of C++'s built-in functions. These functions will be quite useful to you as you continue through this book. Several of them will be used in later chapters. It would be well worth your time to ensure that you are well-acquainted with each of these functions.

MATH OPERATIONS

C++ has a rich set of mathematical operations that are at your disposal. Learning to use these will be quite useful to you as we move throughout this book. To utilize these functions you need to include the math header file.

```
#include <cmath>
```

This header file includes a lot of math functions that are very useful. The most commonly used functions are listed in Table 4.2.

TABLE 4.2   Math Functions

Function	Purpose
double cos(double);	This function takes an angle (as a double) and returns the cosine.
double sin(double);	This function takes an angle (as a double) and returns the sine.
double tan(double);	This function takes an angle (as a double) and returns the tangent.
double log(double);	This function takes a number and returns the natural log of that number.
double pow(double, double);	With this function, you pass it two numbers. The first is a number you wish to raise and the second is the power you wish to raise it to.
double hypot(double, double);	If you pass this function the length of two sides of a right triangle, it will return you the length of the hypotenuse.
double sqrt(double);	You pass this function a number and it gives you this square root.
int abs(int);	This function returns the absolute value of an integer that is passed to it.
double fabs(double)	This function returns the absolute value of any decimal number passed to it.
double floor(double)	Finds the integer which is less than or equal to the argument passed to it.

HINT!

In this table you probably noticed that the function arguments are listed only as types, with no name. When you call the functions you will pass it the name of whatever variable you wish to pass it. What that function internally calls the argument is irrelevant. What is important is remembering what type to pass it and what type it will return.

There are many other math functions available to you when you include this header file, but the functions shown in Table 4.2 are the most commonly used. The following example illustrates some of these functions.

Example 4.8

Step 1: Enter the following code into your favorite text editor.

```
#include <iostream>
#include <cmath>//Math.h is the old C style math
  //header file
using namespace std;
int main()
{
double angle, dsine,dcos,dtan;
cout << "Please enter an angle in radians \n";
cin >> angle;
dcos =cos(angle);
dsine =sin(angle);
dtan =tan(angle);
cout << " Your angles trigonometric functions are
\n";
cout << " Sine is " << dsine << "\n";
cout << " Cosine is " << dcos << "\n";
cout << " Tangent is " << dtan << "\n";
return 0;
    }// main
```

Step 2: Compile the code.
Step 3: When you run the code you should see something like what is shown in Figure 4.8.

You can see that trigonometry is quite easy with C++ math functions. Although not covered here, that math.h header file also has hyperbolic trigonometric functions and arc functions.

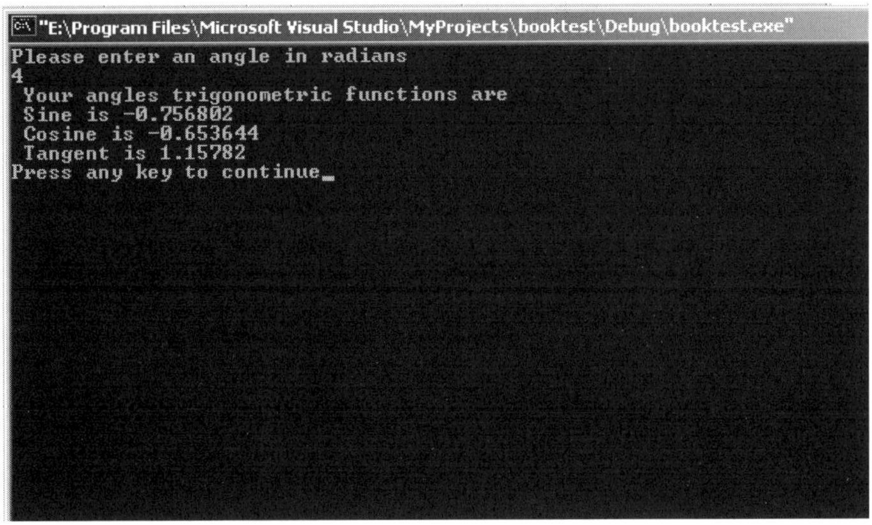

FIGURE 4.8   Math functions.

In addition to trigonometry functions, you have a number of standard math functions at your disposal. You should have noticed some of these listed in Table 4.2. The following example illustrates some of these functions.

Example 4.9

Step 1: Enter the following code into your favorite text editor.

```
#include <iostream>
#include <cmath>
using namespace std;
int main()
{
 double number, dsqrt,dpow,dlog;
 cout << "Please enter a number \n";
 cin >> number;
 dsqrt =sqrt(number);
 dpow =pow(number,5);
 dlog =log(number);
 cout << " Math Example\n";
 cout << " Square Root is " << dsqrt << "\n";
 cout << " Raised to the fifth power is " << dpow <<
```

```
   "\n";
 cout << " Log is " << dlog << "\n";
 return 0;
    }// main
```

Step 2: Compile the code.

Step 3: Run the code. You should see something similar to the image in Figure 4.9.

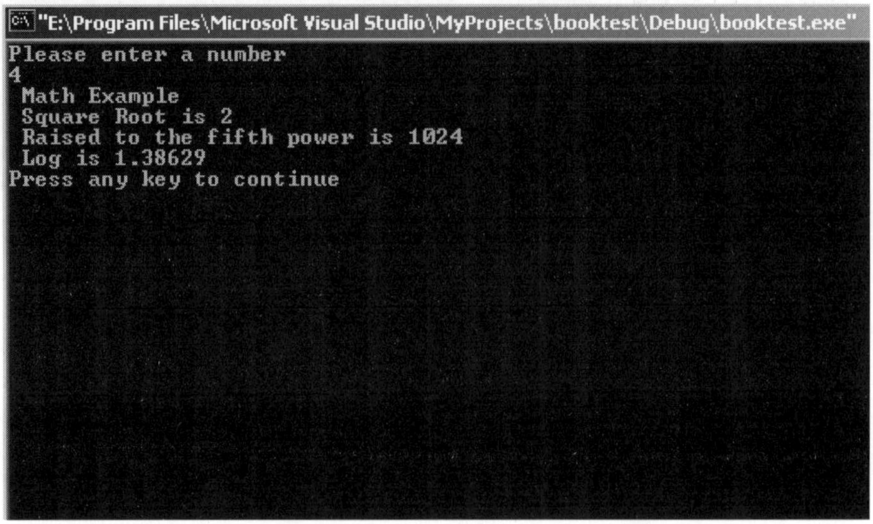

FIGURE 4.9   More math functions.

As you can see, most mathematics is quite simple using C++. It is not uncommon to find that mathematicians who take up programming as a hobby often choose C++ as their language of choice. Chapter 3 demonstrated how easily C++ handles binary numbers, and now you have seen that it has a number of math functions already defined.

TIME
• • • •

It is common for programmers to need to get the time for some reason or another. You might simply wish to display the current time to the user, timestamp a file, or check to see if the current time is within some range

TABLE 4.3   Time Functions Found in Ctime

Function	Purpose
time_t time (time_t * timer);	Gets the number of seconds elapsed since 00:00 hours, Jan 1, 1970 UTC from the system clock.
struct tm * localtime (const time_t * timer);	Converts timer to tm structure, adjusting to the local time zone.
double difftime (time_t timer2, time_t timer1);	Calculates the time difference between timer1 and timer2 in seconds.
struct tm * gmtime (const time_t * timer);	Converts timer to tm structure, adjusting to GMT time zone.
char * ctime (const time_t * timer);	Converts time to a string containing time and date adjusted to local time zone in readable format. This is usually used to present the time in a more readable format.

of dates. C++ provides you with a rich set of tools to do this. You simply need to include the ctime header file and you can access these functions and objects (see Table 4.3).

This is not an exhaustive list, but it does show you the most important functions found in the ctime header file. This might be even clearer for you if you saw an example of this in action. With that in mind, please consider the following example.

Example 4.10

Step 1: Enter the following code into your favorite text editor.

```
#include <iostream>
#include <ctime>
using namespace std;
int main ()
{
time_t rawtime;
time ( &rawtime );
cout << "Current date and time is: "<< ctime
   (&rawtime);
 return 0;
}
```

Step 2: Compile the code and run it. You should see something like what is depicted in Figure 4.10.

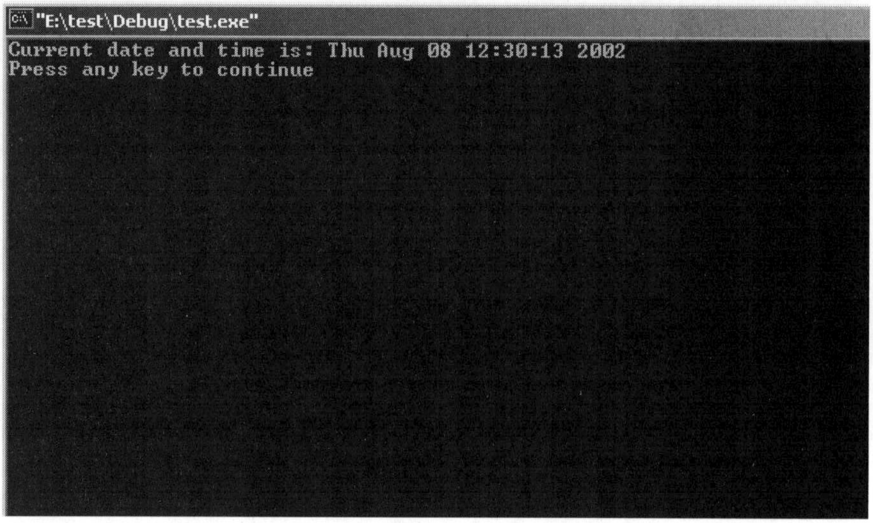

```
"E:\test\Debug\test.exe"
Current date and time is: Thu Aug 08 12:30:13 2002
Press any key to continue
```

FIGURE 4.10   Using time functions.

HINT!

The C method was to include <time.h> and you may still see this occasionally. Most of the functions are the same, at least calling them is the same.

RANDOM NUMBERS

There are many cases where you will wish to generate a random number. If you wished to generate a random number to use as an encryption key, or if you needed random numbers for a game, you would require some function that generates a random number for you.

There are actually two functions you will need to know about. The first is rand(). This function will only return a pseudo random number. The way to fix this is to first call the srand() function. That function seeds the rand() function with a truly random number to start it off. Usually you use the time function to get the number of seconds on your system time, to randomly seed the rand() function. The following example demonstrates this.

Example 4.11

Step 1: Place the following code in your favorite text editor.

```
#include <iostream>
#include <ctime>
using namespace std;
void main( void )
{
 int i,j;
 srand( (unsigned)time( NULL ) );
 /* Display 10 numbers. */
 for( i = 0;i < 10;i++ )
 {
   j= rand();
   cout << j << endl;
     }
}
```

Step 2: Compile and execute the code. You should see something like what is depicted in Figure 4.11

FIGURE 4.11   Using the random functions.

The most important thing to remember is that you must use the srand() function to seed the random function or you will not get truly random numbers. Later in this text when we develop a simple game, you will see the random function used.

SUMMARY

This chapter introduced to you the fundamentals of functions. Functions are the building blocks of any program. In this chapter, you were shown how to create functions, prototype functions, and then call functions. You were also shown how to pass parameters to functions and get return values. At this point, you should be comfortable with creating and using functions in C++.

In addition to the basics of creating and functions, we explored function overloading, parameters, and passing by reference. This material should be sufficient to make you comfortable with creating and using C++ functions. Finally, you were introduced to C++'s rich set of math functions, its time functions, and the random number generation. This should give you a solid understanding of C++ functions.

REVIEW QUESTIONS

1. How do you prototype a function?
2. What are parameters?
3. What are the four rules for writing a function?
4. How many parameters may a function have?
5. What is another word for the parameters of a function?
6. What is function overloading?
7. What is one purpose of header files?
8. What extension do header files end in?
9. Can an overloaded function have the same number of parameters?
10. What is passing by reference?

CHAPTER

5

Decision Structures and Loops

IN THIS CHAPTER

- if Statements
- switch Statements
- for loops
- do loops
- while loops

By now you should have a firm grasp of essentially what C++ is, and know how to create a variable, use functions, and write some simple C++ programs. Now it's time to add to that knowledge, and expand your programming skill-set. This chapter will introduce you to decision structures and loops. These are key concepts that exist in all programming languages; it is just their implementation that is somewhat different. In this chapter, you will learn how to have your code branch in different directions based on certain criteria. You will also learn how to have your code loop repeatedly. A decision structure is a block of code that branches, depending on some decision. There are many times when you will need different code to execute, depending on some criteria such as user input. You have probably used programs in which you were asked to provide some input, maybe a simple *yes* or *no*. The program behaved differently based on the input you gave it. This is an example of a decision structure. Different blocks of code are executed differently based on some criteria. Essentially a decision structure is code that lets your code branch based on some criteria.

Most programs are simply combinations of decisions. If the user does this, then the program responds with one action. If the user does something else, then the program responds with a different action. The entire activity of data processing is dominated by decisions. For this reason, decision structures are a key programming concept found in all programming languages.

IF STATEMENTS

The most common type of decision structure is the `if` statement. These exist in all programming languages but are implemented slightly differently. An `if` statement literally says, "if some condition exists, then do this certain code." A generic, programming-language-neutral example would be the following.

```
If some condition exists
  Execute this code
```

The condition is usually the value of some variable. If a particular variable contains a certain value, or even a range of values, then execute a given block of code. The diagram shown in Figure 5.1A shows the basic structure and flow of an `if` statement.

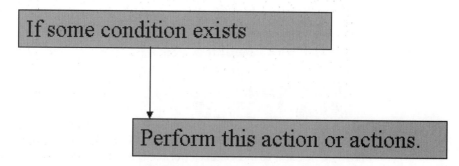

FIGURE 5.1A   The structure of an `if` statement.

It is important to recall that `if` statements are a common facet of almost all programming languages. In every case they will follow the structure and flow shown in Figure 5.1A, but the specific way a particular programming language might go about implementing this concept can be quite different from other languages. However, you should know that many languages use syntax that is similar to the syntax of C++. In fact, `if` statements in C, *Java,* and *JavaScript* are almost indistinguishable from `if` statements in C++. This is important for you to realize. The more thoroughly you master the C++ programming language, the easier it will be for you to learn these other programming languages at some future data, should you so desire.

Let's look at how you implement `if` statements in C++. Here is a basic example that we can dissect to help you learn about `if` statements.

```
If( age == 16)
{
    cout << "Yahoo... Your old enough to drive!\n";
}
```

Let's take a look at this code because there are several things you should note. First of all, you have the word *if* followed by parentheses. Inside the parentheses is where you locate the condition that you base your if-statement on. In other words, if (whatever is in the parentheses is true) do the stuff in the following brackets. The brackets, recall from our discussion of functions, are just boundaries around any block of code. The code in these brackets is what you wish to execute if the statement in the parentheses is true. Finally, notice that there is a double equals sign in the parentheses. Think back to Chapter 1 where you were first introduced to operators. Remember that a single equals is an assignment operator and a double equals is an evaluation operator. By using the double equals, you are asking "is the age equal to 16?" If you used the single equals, you would be stating, "make the age equal to 16."

WATCHOUT!
. . . .

One of the most common mistakes that beginners make is to use a single equals (assignment operator) in an `if` statement rather than a double equals. If you use the single equals—the assignment operator instead of the double equals—the equality operator—you will make the `if` statement true every time.

Example 5.1

Let's use an if statement in a simple program. What we are going to do is to use a simple menu displayed on the screen, and then proceed based on what the user selects.

Step 1: Type the following code into your favorite text editor. Save it as *05-01.cpp.*

```cpp
#include <iostream>
using namespace std;
float square_num(float num);
float cube_num(float num);
int main()
{
  int choice;
  float number, answer;
  cout << "Would you like to (1) square a number or
  (2)
cube a number\? \n";
  cout << "Please press the number of your choice 1 or
  2 \n";
  cin >> choice;
  cout << "Please enter your number \n";
  cin >> number;
  if (choice == 1 )
  {
answer = square_num(number);
  }
  else
  {
answer = cube_num(number);
  }
cout << "Your answer is " << answer <<" \n";
return 0;
}
float square_num(float num)
{
  float a;
  a = num * num;
  return a;
}

float cube_num(float num)
{
```

```
    float a;
    a = num * num * num;
    return a;
}
```

Step 2: Compile your code.

Step 3: Run your code. You should see something like the image shown in Figure 5.1B.

FIGURE 5.1B   if Statements.

What you see here is the use of an if-else statement. If some condition is true, execute the first block of code. If that condition is not true, then execute the second block of code. This is a very common programming situation, and if statements exist in all programming languages, although their implementations may differ. if statements are a funda-mental part of programming. Most programming tasks can, at some level, be reduced to "if some condition exists do this."

It is possible to have if statements and else statements without enclosing the blocks of code with brackets, if and only if the code block consists of one line of code. Consider the following examples.

```
    if(x == 7)
    y++;
```

This is perfectly legal. The C++ compiler will assume that the very first line following the `if` statement is the block of code to execute, even without brackets. However the following code segment is not OK.

```
if (x==7)
    y++;
    z-;
```

In this case, only the first line, the y++, will be executed based on the `if` statement. The second line will not be considered part of the `if` statement code block. In other words, the second line is going to execute regardless of whether the `if` statement is true.

You should also note that although all these examples have used equivalence (if the variable equals some value), you can also use other comparisons, besides equivalence. You can use *greater than* and *less than* comparisons as well. The following code fragments are perfectly valid C++ `if` statements.

```
if (x > 6)
if (j < 10)
```

You can also use *not equal, equal or greater than*, and *equal or less than* as you see in the following examples.

```
if (x !=6)
if (j =< 10)
```

Other possibilities include ranges of values. For example, you may want to know if a particular variable is between two values such as in the following example.

```
if (5 < j < 10)
```

Asks if j is greater than 5 but less than 10. This particular situation is quite common.

`if` statements are also the most common place to see logical AND (&&) and logical OR (||) used.

```
If(j ==5 && I ==6)
... if(j < 4 || j> 10)
```

You should recall that the logical AND and logical OR were both briefly introduced in Chapter 1, in the section on operators. The following example shows the use of these operators in `if` statements.

Example 5.2

Step 1: Type the following code into your favorite text editor.

```
// Include statements
#include <iostream>
using namespace std;
float cube_number(float num);
float square_number(float num);
int main()
{
float number;
float number3;
cout << "Please enter a number \n";
cin >> number;

if (number > 0 && number < 100)
{
  number3 = cube_number(number);
  cout << number << "cubed is "<< number3;
}// end of if
else
{
   number3 = square_number(number);
   cout << number << "squared is "<< number3;
}// end of else
if (number3 <10000 || number3 ==2)
{
   number3 = square_number(number3);
   cout << number3 << "squared is "<< number3;
}// end of second if
return 0;
}// end of main
float square_number(float num)
{
 float answer;
 answer = num * num;
 return answer;
```

```
  }
  float cube_number(float num)
  {
   float answer;
   answer = num * num * num;
   return answer;
  }
```

Step 2: Compile the code.

You should see that you can easily use logical OR and logical AND in your `if` statements.

This gives you a versatile set of comparisons you can use in your code to create `if` statements. You will find, both in this book and in practical experience, that `if` statements are very common in most programming, regardless of the programming language used.

Another possible permutation of the `if` statement is the nested `if` statement. This is having one `if` statement inside another. This is not complicated, and if you indent your code properly it will be fairly easy to follow. The following is an example.

```
if (age >16)
{
if (validlicense == true)
{
    cout << "Yes you can borrow the car \n";
}// end of the inner if
}// end of the outer if
```

The thing to realize about this code is that the inner `if` statement will never be checked if the outer `if` statement is false. This is an example of a nested `if` statement. One `if` statement is nested within another. You should be careful of nesting too many `if` statements. It eventually makes for code that is difficult to read and follow.

SWITCH STATEMENTS

`if` statements and `if-else` statements are great if you have one or two choices. But what happens when you have several choices? You can still use a series of `if-else` statements, but that quickly becomes cumbersome. Fortunately, there is another choice—the `switch` statement. A `switch` statement is literally saying, "switch the path you take through the code based on the value of some variable." A switch can only be done on a single character or on an integer value. You can also think of a switch statement as a complex version of the if statement. It's a way of using if statements when there are multiple possible choices. switch statements, like if statements, exist in most programming languages and the structure is the same. The structure for a switch statement is shown in Figure 5.1C.

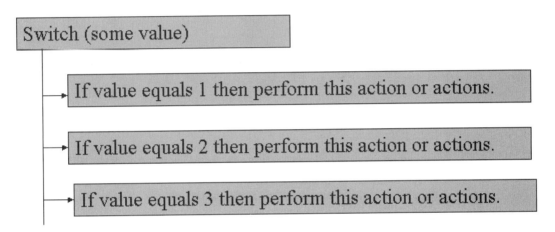

FIGURE 5.1C   Structure of a switch statement.

The implementation of the switch statement in C++ is relatively straightforward. Let's take a look at an example in the following code segment.

```
int choice
switch (choice)
{
case 1:
cout << "You chose 1 \n";
break;
case 2:
cout << "You chose 2 \n";
break;
  case 3:
cout << "You chose 3 \n";
break;
  default:
cout << "You made an invalid choice \n";
```

```
   break;
   }
```

If you examine this code, you will notice several things. First, you will notice the brackets. The brackets start right after the switch statement and end after the last statement in the switch. A switch statement is a code block, just like an entire function or an if statement. And remember that all code blocks are bounded by brackets. You should also notice that each case has a number then a colon. What the code is literally saying is "if it is the case that the integer choice is equal to one, then execute this code." At the end of each case, there is a break statement. This is very important. Without it, the code will continue to execute and will run the next case statement as well. Finally, we have a default statement. If the variable we are switching on does not match any of our cases, then the default will occur.

It is sometimes helpful to think of a switch statement like a switch on a railroad track. You will switch the direction you travel through the code, based on the value of the variable you have selected to base your switch on. Like if statements, the switch is common to most programming languages, just its implementation is different. Lets take a look at a program that uses a switch case statement to create a menu the user can select from. This program will also incorporate other concepts that we have already covered in this book.

Example 5.3a

Step 1: Open your favorite text editor and type in the following code, and save it as *05-03a.h*.

```
void menu();
float circle_area(float);
float triangle_area(float, float);
float rectangle_area (float, float);
```

Step 2: Open your favorite text editor and type in the following code.

```
#include "05-03.h"
#include <iostream>
using namespace std;
int main()
{
   // call the menu to cause it to display initially
```

```
 menu();
 return 0;
}// end of main function
void menu()
{
 int menuchoice;
 float radius, height, width, base, answer;
 char somechar;
// This first section simply displays the users choices
// then
// prompts the user to make a selection.
cout << "1. Area of a circle \n";
cout << "2. Area of a triangle \n";
cout << "3. Area of a rectangle \n";
cout << "4. Exit \n";
cout << " \n  Please enter the number of your selection
 \n";
cin >> menuchoice;   // whatever choice the user makes
// will be stored in the variable
// menuchoice
 // This next section uses a switch case statement to
 //  determine the course of action
 // based on what the user chose.
switch (menuchoice)
{
case 1: // area of circle
 cout << "Please enter the radius of the
circle \n";
 cin >> radius;
 answer = circle_area(radius);
 break;
case 2: //area of triangle
 cout << "Please enter the base of the
   triangle \n";
 cin >> base;
 cout << "Please enter the height of the
   triangle \n";
 cin >> height;
 answer = triangle_area(base, height);
 break;
case 3: // area of rectangle
 cout << "Please enter the height of the
  rectangle \n";
 cin >> height;
```

```
cout << "Please enter the width of the
 rectangle \n";
cin >> width;
answer = rectangle_area(height, width);
break;

  case 4: // exit
return;
  default:
  cout << "Sorry, that was not a valid entry.
  Please try again \n";
  menu(); // it redisplays the menu
 }// end of switch
 cout << "The answer is " << answer << "\n";
 cout << " Press the enter/return key to continue
\n";
 cin >> somechar ; // it does not matter what they
// press
 menu () ; // redisplay the  menu.
}// end of menu function
float circle_area(float radius)
{
 float area;
 area = 3.14 * ( radius * radius);
 return area;
}
float triangle_area(float height, float base)
{
  float area;
  area = .5 * (base * height);
  return area;
}

float rectangle_area(float height, float width)
{
  float area;
  area = height * width;
  return area;
  }
```

Step 3: Compile the code.

Step 4: If all goes well, you should see something like what is displayed in Figure 5.2A.

```
Command Prompt (2) - 04-02
:\Borland\BCC55\Bin>04-02
. Area of a circle
. Area of a triangle
. Area of a rectangle
. Exit

 Please enter the number of your selection

lease enter the radius of the circle

The answer is 50.24
Press the enter/return key to continue
. Area of a circle
. Area of a triangle
. Area of a rectangle
. Exit

 Please enter the number of your selection

lease enter the height of the rectangle

lease enter the width of the rectangle

The answer is 20
Press the enter/return key to continue
. Area of a circle
. Area of a triangle
. Area of a rectangle
. Exit

 Please enter the number of your selection
```

FIGURE 5.2A   Switch statements.

HINT!
• •

You should note that when we include a header file that we create, it is put inside quotation marks, not inside brackets. Header files that are not found in the include directory that installed with your compiler must be put inside quotation marks, and must include the *.h* extension. The *include* directory is the directory/folder where the compiler looks for various header files such as `iostream`, `fstream`, and `cmath`.

This is the longest program you have seen yet in this book but don't be concerned. We are going to review this program, piece by piece, and see what is happening. Hopefully your reading thus far, combined with the numerous comments in the previous code, have given you at least some understanding of what is happening. To begin with, we put the menu in a separate function. The reason for this is so that you could call it repeatedly. You will note that after the answer is returned and displayed on the screen, that we call the menu function again. This means that each time an operation is done, the menu will redisplay and the user can then choose to either perform another operation, or to exit. You should also notice that the function prototyping is all done in a header file.

The actual menu function is the largest and most complex function in this program. The first part of it simply uses a series of cout statements to display choices on the screen, then it prompts the user to make a selection. That selection is stored in a variable named menu-choice. (This name indicates what the variable does.) Then we have a switch case statement that selects a course of action based on the value in that variable. Depending on what choice the user made, the user would be prompted for specific input, the area of the shape they chose would be computed, and then the answer would be returned. Because returning the answer will be the same for all the various choices, that is left until after the switch case. Also notice that after the answer is displayed, the menu is redisplayed. This allows the user to make multiple selections.

WATCHOUT!

Remember that you must have break statements after each case. Without them, the next case statement will also be executed.

As you can see, the switch statement is more versatile than the if statement. Anytime you have more than two or three choices you should definitely consider using the switch statement rather than an if statement. Also note you have a default option with switch statements that you do not have with if statements.

It is important to realize that each case statement can contain any valid C++ code. That might be a series of statements, a call to a function, or even an if statement or another switch statement. You can nest switch statements with other switch statements, if statements, or even loops, just as you nested if statements. The following example illustrates this.

Example 5.3b

Step 1: Type the following code into your favorite text editor and save it as *5.3b.cpp.*

```cpp
#include <iostream>
using namespace std;
void menu();
int main()
{
 menu();
 return 0;
}// end of main
void menu()
{
int choice;
     float area, base, height, radius;
     cout << "1. Area of triangle \n";
     cout << "2. Area of circle \n";
     cout << "3. Exit \n";
     cin >> choice;
     switch(choice)
     {
     case 1:
cout << "1. metric \n";
     cout << "2. english \n";
     cin >> choice;
     cout << "Enter base \n";
     cin >> base;
     cout << "Enter Height \n";
     cin >> height;
     area = .5f * (base * height);
     switch(choice)
     {
     case 1:
    cout << "The area is "<<area << " square
    meters \n";
      break;
      case 2:
    cout << "The area is "<< area <<" square feet
   \n";
      break;
```

```
        default:
        cout << "Invalid Choice \n";
        }// end of inner switch
        break;
        case 2:
        cout << "1. Metric \n";
        cout << "2. English \n";
        cin >> choice;
        cout << "Enter the radius \n";
        cin >> radius;
        area = 3.14f * (radius * radius);
        switch(choice)
        {
        case 1:
      cout << "The area is "<<area << " square
    meters \n";
        break;
        case 2:
      cout << "The area is "<< area <<" square feet
  \n";
        break;
        default:
        cout << "Invalid Choice \n";
        }// end of inner switch

        break;
        case 3:
  return;
        break;
        default:
        cout << "Invalid selection \n";
        }
        menu();
  }
```

Step 2: Compile and execute this code. You should see something similar to what is shown in Figure 5.2B.

You can see how nested `switch` statements can be used to provide another layer of possible branching. Nested `if` statements, `switch` statements, and loops are very common, you will probably encounter them often.

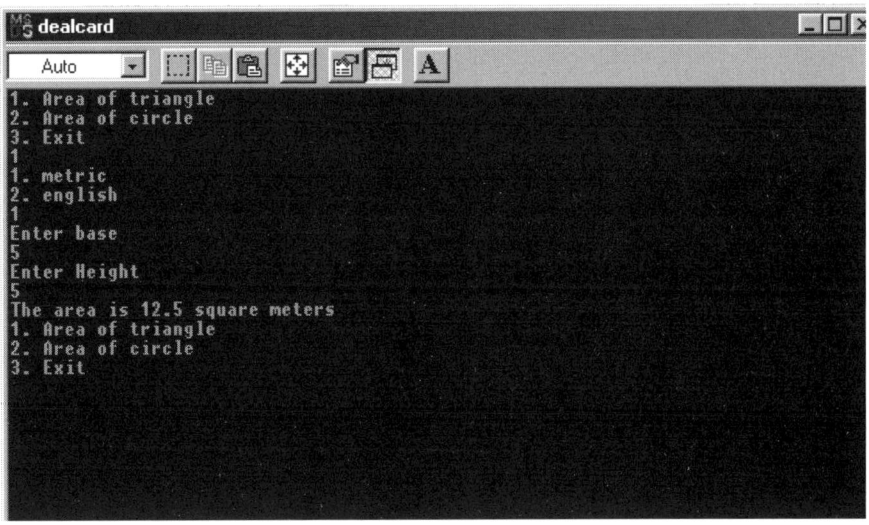

FIGURE 5.2B   Nested switch statements.

FOR LOOPS

The for loop is perhaps the most commonly encountered loop in all programming. Its structure is the same in all programming languages, only its implementation is different. It is quite useful when you need to repeatedly execute a code segment, a certain number of times. The concept is simple. You tell the loop where to start, when to stop, and how much to increase its counter by each loop. The basic structure of a for loop is shown in Figure 5.2C.

The essence of the for loop is in the counter variable. It is set to some initial value, it is incremented (or decremented) each time the loop executes. Either before or after each execution of the loop the counter is checked. If it reaches the cut-off value, stop running the loop.

The specific implementation of the for loop in C++ is shown here:

```
for (int i = 0; i < 5; i++)
{
}
```

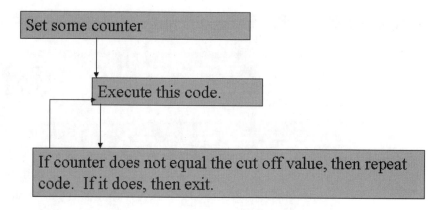

FIGURE 5.2C   The structure of a for loop.

What this code is doing is rather straightforward. First, we create an integer to be used as a loop counter (to count how many times the loop has executed) and we give it an initial value of zero. We then set the conditions under which the loop should execute. That's the i < 5 portion of the code. It is essentially saying, "keep going as long as i is less than 5."As soon as that condition is no longer true (i.e., i is equal to 5 or greater than 5), the loop stops executing. Finally, we have a statement to increment the loop counter after each iteration of the loop. Notice the semicolons in the for loop declaration. You should recall that all statements/expressions in C++ end with a semicolon. These code snippets are, indeed, valid statements and could be written as stand-alone expressions. The reason the last one does not have a semicolon is because it is followed by a closing parentheses and that terminates the expression. The three parts to the for loop declaration are the following.

1. Declare and initialize the loop counter.
2. Set the conditions under which the loop will execute.
3. Increment the loop counter.

Example 5.4

Step 1: Use your favorite text editor to enter the following code.

```cpp
#include <iostream>
using namespace std;
int main()
{
 for (int j = 0; j < 10; j++)
 {
cout << "Going Up…." << j << "\n";
 }// end of for loop
 return 0;
}// end of main
```

Step 2: Compile the code.

Step 3: When you run the code, you will see something similar to what is displayed in Figure 5.3.

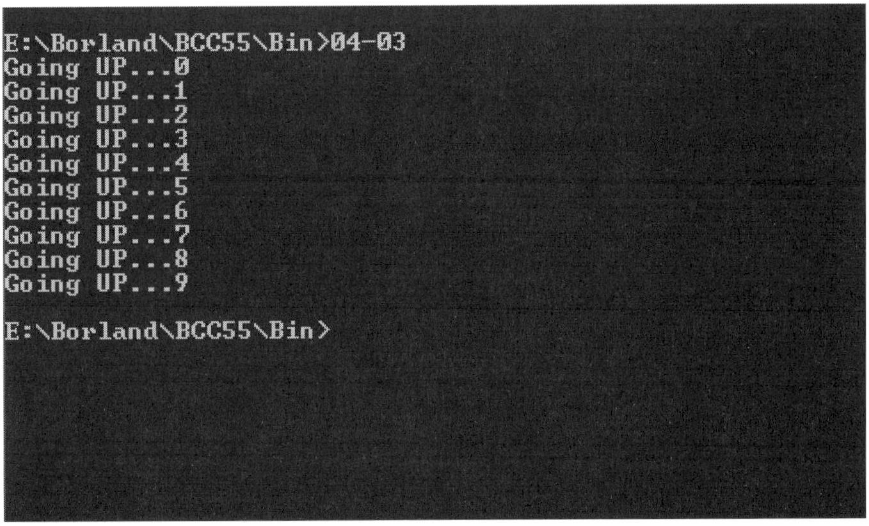

```
E:\Borland\BCC55\Bin>04-03
Going UP...0
Going UP...1
Going UP...2
Going UP...3
Going UP...4
Going UP...5
Going UP...6
Going UP...7
Going UP...8
Going UP...9

E:\Borland\BCC55\Bin>
```

FIGURE 5.3   For Loops.

This simple example will show you how a `for` loop works. Its rather simple but should give you the basic idea.

It should be pointed out that it is possible to increment by values greater than one, and even to count backwards, using the decrement operator. The next example illustrates this:

Example 5.4a

Step 1:

```
#include <iostream>
using namespace std;
int main()
{
 for (int j = 10; j >0; j—)
 {
cout << "count down… << j << "\n";
 }// end of for loop
 cout << "Blast Off!!! \n";
}// end of main
```

Step 2: Compile the code.

Step 3: When you run this program you should see something like what is displayed in Figure 5.4A.

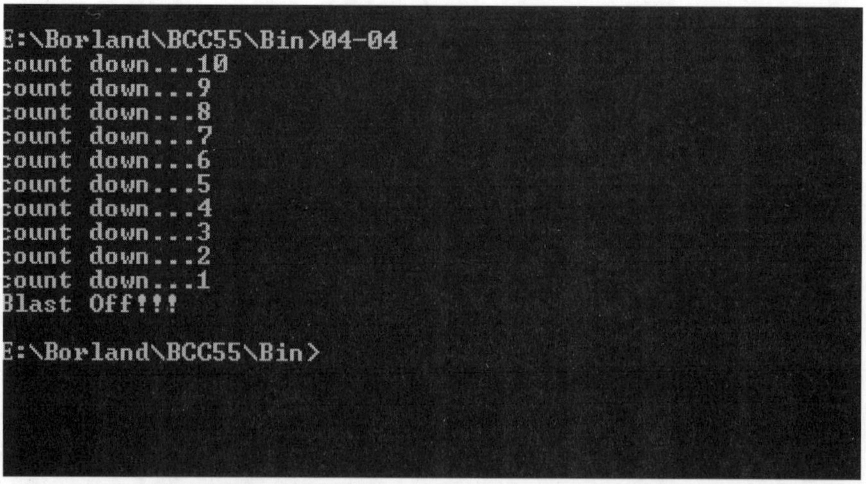

FIGURE 5.4A For loops 2.

You will find many programming situations where the `for loop` is quite useful, so it's rather important. Also remember that `for loops`, like `if` statements, are found in all programming languages, only their implementation is different.

Loops, like `if` statements and `switch` statements, can be nested. You can have a loop inside of another loop. When you study algorithms in Chapter 13, and games in Chapter 14, you will see places where this is practically applied. The most important thing to remember when nesting loops is that the inner loop will execute a number of times equal to its counter multiplied by the outer loops counter. Put another way, if your outer loop executes 5 times, and you set the inner loop to loop 4 times, the inner loop will actually loop 20 times. It will loop 4 times, for each iteration of the outer loop. Let's look at an example of this.

Example 5.4b

Step 1: Enter the following code into a text editor and save it as *05-04b.cpp*.

```
#include <iostream>
using namespace std;
int main()
{
for (int i = 0;i<3;i++)
{
   for(int j=0;j<4;j++)
      {
cout << "This is inner loop " <<  j ;
     cout << " of outer loop " << i << endl;
        }// end of inner loop
 }// end of outer loop
 return 0;
   }// end of main
```

Step 2: Compile and execute the code. You should see something like what is displayed in Figure 5.4B.

Note how often the inner loop printed out. This should prove to you the earlier statement regarding the number of iterations an outer loop makes. The outer loop, in this case, was set to loop 3 times. The inner loop was set to loop 4 times. Because the inner loop was nested within the outer loop, it actually looped 3 multiplied by 4, or 12 times.

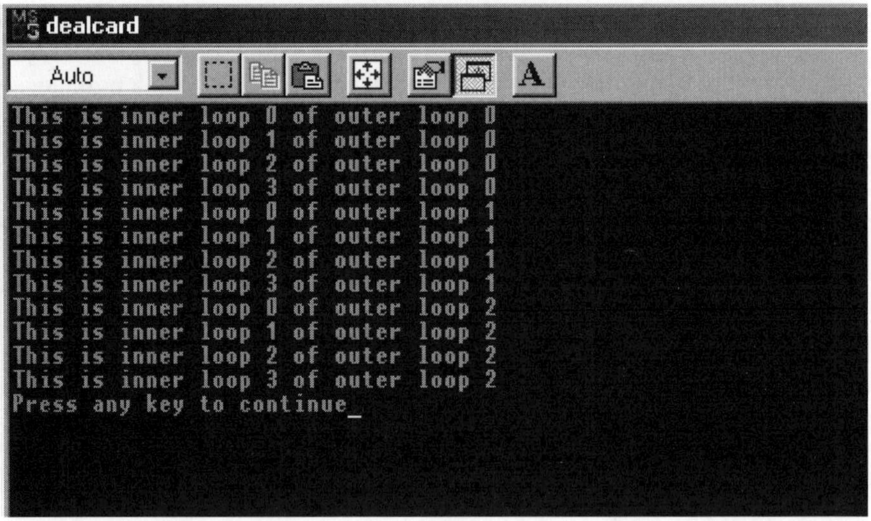

FIGURE 5.4B Nested for loops.

DO LOOPS

The for loop is the most common type of loop; however, it is not the only type of loop. There are other loop structures you can use, although they all accomplish the basic goal of executing a given piece of code a specified number of times. The do loop is one such loop structure. Unlike the for loop, however, it does not evaluate whether or not the loop should continue until after the loop has completed an iteration. The following is the generic structure.

```
Do
{
  // place code here
} while (some condition is true)
```

The following is a more specific example.

Example 5.5

Step 1:

```
#include <iostream>
using namespace std;
```

```
int main()
   {
   int x;
  do
  {
cout<< x  ;
x++;
  }while (x < 10);
  return 0;
   }
```

You can see that the do loop does essentially the same thing as the for loop. However, the loop condition is not evaluated until after the loop has executed an iteration, and the loop counter is incremented inside the loop. These loop structures give you different options for accomplishing the same basic goal.

WHILE LOOPS

While loops are almost identical to do loops. They are simply structured a little bit differently. The do loop has the condition to be evaluated at the end of the loop, whereas the while loop has it at the beginning.

```
while(condition)
{
}
```

The following is an example of a while loop.

Example 5.6

Step 1: Type the following code into your favorite text editor.

```
#include <iostream>
using namespace std;
int main()
{
 int i = 0;
 while(i < 10)
 {
    cout << "I is " << i << "\n";
```

```
        i++;
     }
     return 0;
}// end of main
```

Step 2: Compile the code.

Step 3: Run the code. You should see something like the image in Figure 5.5.

FIGURE 5.5 While loops.

As you can see, all loop structures accomplish essentially the same goal. They execute a given block of code a certain number of times. The only differences are where you check the value of the loop to see if you will continue looping, and where you increment the loop counter. C++ loops are best described by the old axiom "There is more than one way to skin a cat."

SUMMARY

This chapter has added a vital component to your programming knowledge. It has included the topics of how to make decisions in your code using either an `if` statement or a `switch case` statement to choose different tracks through your program. These decision-making structures are vital to any program and are present in all programming languages. You should now be able to use a `for loop` and understand its structure.

This chapter also introduced you to alternative loop structures such as the do loop and the while loop.

REVIEW QUESTIONS

1. What are the parts to a for loop declaration?
2. What types of variables can be used for a switch statement?
3. What is the purpose of the default statement in a switch statement?
4. Why is it important to use the double equals sign in an if statement?
5. Why might a switch statement be better than an if statement in some cases?
6. What happens if you omit the break statements from your switch block?
7. What is the purpose of a for loop?
8. Can you have an if statement without enclosing brackets?
9. How do you implement more than one choice in an if statement?
10. What other possibilities besides equivalence can you use for if statements?

File Input and Output

Chapter 2 introduced you to screen input and output. Chapters 3, 4, and 5 made use of this practical information. In this chapter, you will learn how to get input from a flat file and send output to a flat file. This can be very useful. There are times when you will need to import data from a comma delimited text file, or perhaps you wish to write errors to a log file. All this, and more, can be accomplished with file input and output.

Importing data from a flat file is, in essence, quite similar to getting data in from the keyboard. And outputting to a file is similar to outputting data to a screen. C++ handles both in similar ways, so if you fully understood the input and output techniques in Chapter 2, then this chapter should not pose any great challenges for you.

WHAT IS A FLAT FILE?

A flat file is simply a file that has no structure. Thus, its name—it is flat, in that it has no structure or relationships. Data is simply put into the file with no interrelationships between entries in a file, or between files. Database files (such as *Microsoft Access* files or *.mdb* files) have a definite structure. Database files also have clearly defined relationships between entries (called records) and even between different files. However, long before there were relational databases, there where flat files. You might think that with the advent of relational databases flat files are now irrelevant. However, this is not true. There are many cases where it is more appropriate to store data in a flat file than in a complex relational database. One such situation is when writing a log of events or errors.

When you open *notepad* and create a text file, you are creating a flat file. Any file that simply contains ASCII (The American Standard Code for Information Interchange), pronounced AZ KEY, text is a flat file. In addition to ASCII code there is now Unicode. Unicode is simply an expanded version of ASCII, made to handle international symbols. In fact, the first 255 characters of the Unicode set are the ASCII codes. An in-depth discussion of ASCII and Unicode is beyond the scope of this book. Just remember that text files are flat files. And remember what the ASCII character codes are. Figures 6.1A and 6.1B show the entire ASCII code set. You can also find references to the ASCII code set on the Internet (search ASCII character code) and in the help files that ship with many commercial software development tools.

It is not important to memorize these character codes, but you will probably refer back to this table frequently. The ASCII codes are used again in this book in a number of places, including in games programming. You should notice, however, that there exists an ASCII code for every keystroke available on your keyboard, as well as some basic symbols. The ASCII character code set is widely used throughout programming, and can be used in most programming languages.

0	☐	32	[space]	64	@	96	`
1	☐	33	!	65	A	97	a
2	☐	34	"	66	B	98	b
3	☐	35	#	67	C	99	c
4	☐	36	$	68	D	100	d
5	☐	37	%	69	E	101	e
6	☐	38	&	70	F	102	f
7	☐	39	'	71	G	103	g
8	* *	40	(	72	H	104	h
9	* *	41	)	73	I	105	i
10	* *	42	*	74	J	106	j
11	☐	43	+	75	K	107	k
12	☐	44	,	76	L	108	l
13	* *	45	-	77	M	109	m
14	☐	46	.	78	N	110	n
15	☐	47	/	79	O	111	o
16	☐	48	0	80	P	112	p
17	☐	49	1	81	Q	113	q
18	☐	50	2	82	R	114	r
19	☐	51	3	83	S	115	s
20	☐	52	4	84	T	116	t
21	☐	53	5	85	U	117	u
22	☐	54	6	86	V	118	v
23	☐	55	7	87	W	119	w

FIGURE 6.1A   The first 126 characters of the ASCII code set.

128	€	160	[space]	192	À	224	à
129	€	161	¡	193	Á	225	á
130	€	162	¢	194	Â	226	â
131	€	163	£	195	Ã	227	ã
132	€	164	¤	196	Ä	228	ä
133	€	165	¥	197	Å	229	å
134	€	166	¦	198	Æ	230	æ
135	€	167	§	199	Ç	231	ç
136	€	168	¨	200	È	232	è
137	€	169	©	201	É	233	é
138	€	170	ª	202	Ê	234	ê
139	€	171	«	203	Ë	235	ë
140	€	172	¬	204	Ì	236	ì
141	€	173		205	Í	237	í
142	€	174	®	206	Î	238	î
143	€	175	¯	207	Ï	239	ï
144	€	176	°	208	Ð	240	ð
145	€	177	±	209	Ñ	241	ñ
146	€	178	²	210	Ò	242	ò
147	€	179	³	211	Ó	243	ó
148	€	180	´	212	Ô	244	ô
149	€	181	µ	213	Õ	245	õ
150	€	182	¶	214	Ö	246	ö
151	€	183	·	215	×	247	÷

FIGURE 6.1B   The last half of the ASCII code set.

IFSTREAM AND FLAT FILES

The `ifstream` object is used to input from files. For now, you should think of an object or a class as a group of related functions all grouped together into a single bundle. That bundle is what we are calling a class or an object. We will be exploring in greater detail exactly what classes and objects are later in this book. In fact, three chapters are devoted to this topic.

The word *stream* in the name should be of interest to you. Files are input and output in C++ in the form of a stream. A stream is simply a running thread in memory. It is literally a stream of bytes, with each byte representing a character in the file. In the C++ programming language, all input and output occurs in streams of bytes. A stream is simply a sequence of bytes. Input streams receive bytes from some device—usually the keyboard or, in the case of `ifstream`, a file on the disk.

Basically, the `ifstream` class is a group of functions used to input files. The `ofstream` class is used to output to files. The functionality of both are combined in `fstream`. These objects can be used in your code to represent specific files, and to manipulate those files. You can input and output to a file using `ifstream`, `ofstream`, and `fstream`. Each of these objects will treat the file as a stream of bytes, with each byte representing a single character in the file. Thus, this is the reason for the word *stream* in the names of these objects. It is the same as with `cout` and `cin` from the `iostream` header file. They treat the keyboard input and the screen output as a stream of bytes. The only real difference is that `cout` sends data to the screen and `ofstream` sends it to a flat file. They both handle the data in much the same manner.

Output streams send their sequence of bytes to some device—usually the monitor—but in the case of `ofstream` the bytes are sent to some file on the hard disk. As was mentioned earlier, the underlying function is the same as output you saw in Chapter 2, it is only the destination that is really different.

You can create an instance of this class just like you would any other variable with one exception. After the name you give your variable you will have open and close parentheses. In some cases, these parentheses will have parameters in them; in others they will not. The following is the most common, and the simplest, way to create an instance of this `ifstream`.

```
ifstream fin();
```

With this line of code, we have just created a variable called `fin`. That variable is of the type `ifstream`. That means that it is an object that represents a file to be input. This object, `ifstream`, has a number of methods and properties that you will be able to use in your programs. Table 6.1 summarizes the most commonly used of these.

You can see that a lot of these methods mention a *stream*. You might be wondering what this is referring to. The most important thing for you to remember is that C++ handles file input and output as streams of characters. Actually, it handles screen/keyboard input and output the same way, but with `cout` and `cin` it's not as obvious. A stream is simply a flow of individual characters, flowing much like a stream of water, thus the name.

The properties are designed simply to let you know the current state of the file. (Is it open or not?) The various methods you see in Table 6.1 are primarily concerned with opening the file, reading it in (either one character at a time, or one line at a time), and closing the file. The exception is

TABLE 6.1   `ifstream` Methods and Properties

Method/Property	Purpose
clear	This method simply clears the contents of the buffer.
close	Closes the file you previously opened.
eof	This property is a Boolean and tells you if you are at the end of the file you have opened.
get	This method gets a single character.
getline	This method gets an entire line.
is_open	This is a Boolean property that is true if the file is open and false if it is closed.
open	Opens a file and attaches it to the filebuf object and, thus, to the stream.
setmode	Sets the stream's mode to binary or text.
read	This method simply reads in data from the stream. The data will be in binary mode.
seekg	When a file is opened in binary mode you can tell it to go to a particular space in the file. This method does that.
tellg	This method retrieves the current position of the pointer in the file.

the `setmode` method. It determines how the file will be opened. By default it is opened as text. You do, however, have some options as to how you open it. You can set the mode to text or binary. You will probably use the text mode most of the time.

Example 6.1

This example will illustrate how to read-in text data from a text file. The data is simply read in to a variable then displayed on the screen.

Step 1: Enter the following code into your favorite text editor.

```
#include <iostream>
   #include <fstream>
using namespace std;
int main ()
   {
 char buffer[256];
 ifstream myfile ("test.txt");
 while (! myfile.eof() )
 {
   myfile.getline (buffer,100);
   cout << buffer << endl;
 }
 return 0;
}
```

Step 2: Compile the code.
Step 3: Run the code. You should see something like what is shown in Figure 6.1C.

FIGURE 6.1C   File input.

Let's consider this code for a few moments to make sure you understand what is happening. First, we have the include statements that you see at the beginning of all C++ programs. The fstream file is where we get the file input stream that allows us to read in a file. Next, inside the main function, we declare a character array named buffer. This is literally a buffer that is used to store the data as it's read in.

The next line of code that is of interest to us is the while (!myfile.eof()). This is a basic while loop that is using the negation operator (!) on the end of file property of the myfile object. What this code is essentially saying is that "while it's not true that this is the end of the file, keep looping." As soon as the end of file is reached, the loop will stop executing. Finally, we see the getline method used to retrieve an entire line of text from the keyboard input. Remember that the getline function allows you to determine how many bytes will be read in, thus preventing you from over flowing an array.

WATCHOUT!

Notice that the file we chose to open in this example has no path, it is in the same directory as the program. If your file is in a different directory, you must use a full path. However, remember that inside strings the \ character denotes an escape key, so to put that character in you have to use a double slash \ \. The following is an example.

```
ifstream myfile ("c:\\myfolder\\test.txt");
```

OFSTREAM AND OUTPUT

You now have a basic understanding of text file input. What about output? You may wish to write things to a file from time to time. It's very similar to inputting a text file. Inputting a file requires the ifstream object. Outputting uses the ofstream object. The methods of ofstream are quite similar to the methods of ifstream with a few exceptions. The major methods are shown in Table 6.2.

This next example opens a file, then writes text to it. All of this is done using the ofstream.

TABLE 6.2   `ofstream` Methods and Properties

Method	Purpose
open	This works just like the open method in ifstream; it opens a file.
close	This method closes a previously opened file.
attach	This attaches your current stream to an existing file. It is useful if you wish to append the contents of your stream to another file.
is_open	This property returns true if the file is open.
seekp	This method sets the pointer location in the file.
tellp	This method retrieves the current pointer in the file.
write	This method writes a sequence of characters to a file.

Example 6.2

Step 1: Open your favorite text editor and enter the following code.

```
#include <fstream>
int main ()
{
 ofstream myfile ("test2.txt");
 if (myfile.is_open())
{
   myfile << "This outputting a line.\n";
   myfile << "Guess what, this is another line.\n";
   myfile.close();
}
return 0;
   }
```

Step 2: Compile the code.

You can see that this code is quite similar to the file input code. Note that the << operator is used here, just as it was with cout. C++ tries to make input and output very similar whether it's screen/keyboard or files. These previous two examples demonstrate the basics of input and output of text files. You will use these techniques again later in this book.

SEQUENTIAL FILES

A sequential file is simply a file where the data is stored one item after another. The flat text files you have dealt with so far are sequential files. A more technical definition would be that a sequential file is a collection of data stored on a disk in a sequential, nonindexed, manner. C++ treats all such files simply as a sequence of bytes. Each file ends with a special end of file marker. Sequential files do not have additional inherent structure. Any additional structure must be imposed by the program reading and writing the file. What this means in plain English is that there is no way to move around in a file. You just start at the beginning and input the entire file from start to finish. You can open these files in a variety of ways. Of course, you will still have to include `ifstream` and/or `ofstream`.

Example 6.3

Step 1: Declare a variable of type `ifstream/ofstream`, depending on whether you need input or output. This variable is often referred to as a file handle.

```
ifstream infile;    // declares file handle called
   // infile
ofstream outfile;   // declares file handle called
   // outfile
fstream inoutfile;  // declares a file handle that
   // can be either
```

Step 2: Open the file for reading, using the ifstream member function `open()`. When you open it you pass it the name/location of the file you wish to open, and the mode you wish to open it in. The mode tells it whether this is input, output, or both.

```
infile.open("myfile.txt", ios::in);  //for reading only
outfile.open("myfile.txt", ios::out); //for writing only
inoutfile.open("myfile",ios::in|ios::out); //for reading
   //& writing
outfile.open("myfile.txt", ios::app); //for appending
outfile.open("myfile.txt", ios::trunc);
```

This last one is rather interesting. The truncate option says that "if the file already exists, then wipe it out and start with a new file." You should, obviously, be quite careful with this one.

When you need to go to file that is not in the same directory/folder as your program, you will need to specify the path.

WATCHOUT!

 Remember that the \ in C++ denotes an escape character. So you must use \\ to denote directories. The following is an example of the wrong way and the right way to do this.
WRONG: infile.open("c:\myfolder\myfile.txt",
ios::in);
CORRECT: infile.open("c:\\myfolder\\myfile.txt",
ios::in);

Let's look at an example that combines all the major modes for opening a file. This will let you see how each of them work.

Example 6.4

Step 1: Enter the following code into your favorite text editor.

```
    #include <fstream>
#include <iostream>
using namespace std;
int main()
{
    char buffer[256];
    // open it for output then write to it
  fstream myfile;
    myfile.open("test2.txt",ios::out | ios::trunc);
  if (myfile.is_open())
  {
myfile << "This outputting a line.\n";
myfile.close();
    }
    // open it for input and read in
    myfile.open("test.txt",ios::in);
    myfile.getline(buffer,100);
    cout << "The file contains   " << buffer << "\n";
  myfile.close();
```

```
//open for appending and append
    myfile.open("test.txt",ios::app);
    myfile << " Hey this is another line \n";
    myfile.close();
    // open for input and print to screen
    // open it for input and read in
    myfile.open("test.txt",ios::in);
    myfile.getline(buffer,200);
    cout << "The file contains   " << buffer << "\n";
myfile.close();
return 0;
}// end of main
```

Step 2: Compile the code.

Step 3: Run the code.

In this single example, you can see a file object being opened several times in several different modes. This should illustrate to you the various uses of the modes for opening a sequential file.

BINARY FILES

Text files are not the only type of flat files you can use. You can also use binary files. Binary files are still flat files; however, they are in basic binary format rather than ASCII. Remember that everything in a computer is binary (1's and 0's). The binary format has each file stored in a byte-by-byte format. This produces some odd-looking results if you try to view binary files like plain text. For example, an integer might be represented by four separate characters, because it occupies four bytes. You might ask why anyone would want to open a binary file. There are a number of reasons, the most prominent being that there is still old data stored this way. However, one useful thing you can do is to open any file in binary mode and count the bytes in it, to see how big the file is. This next example demonstrates this.

Example 6.5

Step 1: Enter the following code in your favorite text editor.

```
#include <iostream>
using namespace std;
#include <fstream>
```

```
int main ()
{
 long start,end;
 // Recall from chapter one, that a long is simply
 // large integers
 ifstream myfile ("test.txt", ios::in|ios::binary);
 start = myfile.tellg();
 myfile.seekg (0, ios::end);
 end = myfile.tellg();
 myfile.close();
 cout << "size of " << "test.txt";
 cout << " is " << (end-start) << " bytes.\n";
 return 0;
}
```

Step 2: Compile this code.
Step 3: Run the executable. You should see something like Figure 6.2.

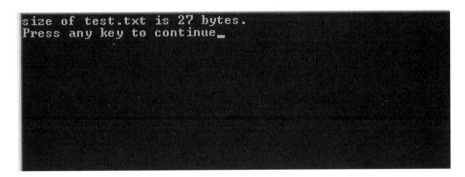

FIGURE 6.2   Binary file size.

This code is relatively straightforward, let's take a look. To begin with, we open a file just as we would normally do, except we open it as binary, with the following line of code.

```
ifstream myfile ("test.txt", ios::in|ios::binary);
```

Next, we retrieve the beginning point for the file.

```
start = myfile.tellg();
```

And then we get the endpoint of the file, by first moving to the end of the file then getting the position.

```
myfile.seekg (0, ios::end);
end = myfile.tellg();
```

The first line says to start at the 0 byte/character and search until the end of the file. If you wanted to start at another point such as the 10th character or byte, you could simply put in a 10 where the 0 is. Now it's just a matter of subtracting the beginning point from the endpoint and that's how many bytes of data the file has in it. The process is not particularly complicated. However, you might be wondering about the `ios` you keep seeing. That is simply C++'s way of saying "input/output stream."

SUMMARY

This chapter has introduced you to the concepts involved in reading and writing from flat files. We have worked through specific examples reading and writing from text files. We have also used an example where we read from a file in binary mode. This chapter summarized the methods and properties of the `ifstream` object and the `ostream` object. You should be fairly comfortable with these topics.

REVIEW QUESTIONS

1. What are two modes you can open a file in?
2. What is the difference between the `get` method and the `getline` method?
3. What header file must you include for file input and output?
4. What is a flat file?
5. What method can you use to change a file's mode?
6. Why would you not want to use the `read` method to read in plain text?
7. What is ASCII?
8. How does the << operator work with file output?
9. What is a binary file?
10. What is the `eof` property of an `fstream` file?

7

Exception Handling

IN THIS CHAPTER
· · · · · · · · · · · · · ·

- Introduction
- Errors and Exceptions
- try-catch Blocks
- Exception Classes
- Throwing Exceptions
- Saving Errors to Logs
- General Testing and Debugging

In any programming situation various problems can occur. You can have basic errors of many types. Handling those problems is of paramount importance. You do not want your program to crash when an error occurs. If this were to occur, the people using your software (programmers generally refer to them as "users") would be very unsatisfied with your product. A better idea is to correct the problem or at least give the user a friendly message telling them what has occurred. Exception handling is, therefore, a critical topic in programming.

Different programmers have different philosophies regarding exception handling. However, it is commonly believed that you cannot have too much exception handling. Code should be contained inside some exception handling.

ERRORS AND EXCEPTIONS

There is an almost limitless list of things that can go wrong. Think back to Chapter 5: What would happen if you tried to open a file that wasn't there? You would get an error, and without error handling your program would crash. What if you asked a user to enter a number and they typed in their name, then your code tried to store their input in an integer variable? You would get an error, and, yes, without error handling your program would crash.

Errors can be grouped into three main categories, each of which is described in Table 7.1.

TABLE 7.1   Categories of Errors

Error Type	Description
Syntax	A syntax error occurs when you have attempted to use the language in a way that is inappropriate. For example if you said "if (x $ $ y)" when you meant to write "if (x==y)" that would be a syntax error. The compiler will not compile your code if you have any syntax errors.
Runtime	This is an error that occurs when you run your program. It is any interruption to the normal flow of your application. Trying to open a file that does not exist and dividing by 0 are two examples of runtime errors. These errors are also called exceptions.
Logic	This occurs when your program compiles and runs properly, but produces erroneous results. There is some error in your programming logic. This can be the hardest type of error to find and is what is generally referred to as a "bug."

Now for the really bad news—you will have errors. It's a fact of life that when you write programs you get errors. The first thing you can do is thoroughly test your programs to attempt to uncover and fix errors before distributing the program. The other thing you can do is provide adequate error handling so that errors don't cause your program to crash. Error handling will take care of runtime errors and the compiler will identify syntax errors. Logic errors are what will be your biggest problem. At the end of this chapter, we will explore some techniques to help you in debugging applications.

TRY-CATCH BLOCKS

C++ offers a rather powerful error-handling tool called the `try-catch` block. This technique is also used in *Java* and *VB.net*™, so this is a skill you will be able to carry over to other programming languages. The `try-catch` block consists of two sections. The first is the code that you want to execute. This is the code that you *try*. If an exception is generated then the `catch` block *catches* it. Thus, the `catch` block is where your error handling goes. The following example illustrates this point.

Example 7.1

Step 1: Enter the following code into your favorite text editor.

```
#include <iostream>
using namespace std;
#include <fstream>
int main ()
{
 try
 {
    char buffer[256];
    ifstream myfile ("test.txt");

    while (! myfile.eof() )
    {
myfile.getline (buffer,100);
cout << buffer << endl;
    }
 }// end of try block
 catch(…)
  {
cout << "There was an error !\n";
  }
 return 0;
}
```

Step 2: Compile the code.

Step 3: Run the code.

Now, if the file is found, then things will work out as they always have. But if for some reason the file is not found, then the runtime error will be caught and the user will get a simple error message. To test this you can simply change the file name to one that does not exist on your PC.

You probably also noticed that the catch block has three dots in the parentheses. This means that all exceptions should be handled by this single catch block. You will see later in this chapter that you can handle different types of exceptions differently.

Essentially, what you see in the previous example is that the code in the try block is attempted. If it executes without error, then the catch block will never be executed. If an error occurs in the try block then execution of that code stops and the catch block is executed. There should be absolutely no code whatsoever between the closing bracket of the try block and the opening bracket of the catch block.

WATCHOUT!
● ●

If you do not use try-catch blocks, then any exception that occurs will cause your program to cease working.

It is also rather common practice for C++ programmers to set up a series of try- catch blocks, each catching a different type of error. What happens is, depending on what is wrong, your code will "throw" a different type of message, such as an int, string, and so on. You then have a series of catch blocks that catch each of these types of errors. Consider the following example.

Example 7.2

Step 1: Enter the following code into your favorite text editor.

```
#include <iostream>
using namespace std;
int main()
{
```

```
int answer, divisor, dividend;
try
{
  cout << "Please enter an integer \n";
  cin >>divisor;
      cout << "Please enter another integer \n";
      cin >> dividend;
      if(dividend ==0)
      throw 0;

      answer = divisor/dividend;
      cout << answer;
      return 0;
}
catch (int i)
{
 cout << "You cannot divide by zero";
}
    }
```

Step 2: Compile the code.

Step 3: Run the code. Enter 0 for the second number. You should see something like what is depicted in Figure 7.1

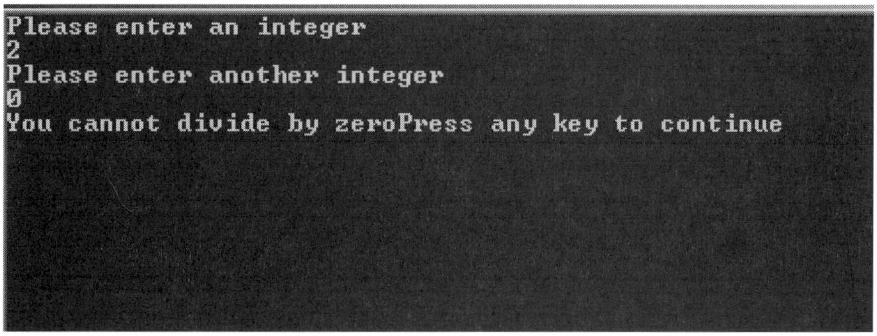

FIGURE 7.1 Multiple catch blocks.

You can see that this catch block only catches integers. If the dividend is 0, then we throw an integer. It would be a simple matter to throw a variety of exceptions using different data types for each error type.

EXCEPTION CLASSES

Up to this point in the chapter, we have handled all the exceptions the same way. This is quick and easy, but not always the best method. Think about it this way—different errors will require different handling. If the error is a "file not found," then you probably want to prompt the user to enter a different file name. If the error is a division by 0 error, then you probably want to prompt the user to enter a non-0 number. Some functions of the standard C++ language library send exceptions that can be captured if we include them within a try block. These exceptions are sent with a class derived from std::exception as type. Each exception has a specific class associated with it. This class (std::exception) is defined in the C++ standard header file <exception> and serves as pattern for the standard hierarchy of exceptions. Again for now you should simply think of a class as a bag of functions. (Classes will be discussed soon!) Figure 7.2 summarizes some of the more commonly encountered exception classes.

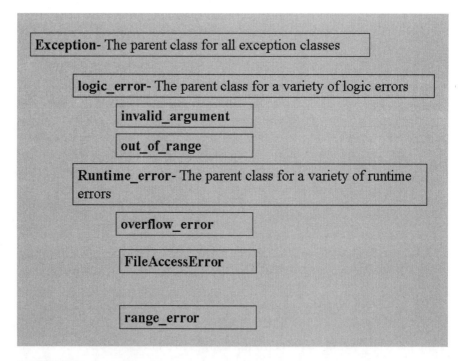

FIGURE 7.2   Exception classes.

HINT!

Microsoft defines its own set of exception classes for use with *Visual* C++, however, the standard exception classes will work with *Visual* C++ and with all other C++ compilers.

The hierarchy you see demonstrated in Figure 7.2 is not exhaustive. It does, however, show you the relationship between exceptions. This becomes important when you are handling multiple exceptions. You can trap specific errors and handle them differently. You must remember that you have to trap exceptions in the reverse order of how they appear on the hierarchy. What this means is that if you trap `run_time_error` you cannot then have another `catch` statement for `FileAccessError` after that. The reason is that the exception higher up in the hierarchy includes all lower exceptions that come under it in the hierarchy. Put another way, you should trap the most specific and, therefore, lower-level exceptions first, then work your way toward more general exceptions.

Once we move forward into classes and object-oriented programming, you will learn how to create your own exception objects.

THROWING EXCEPTIONS

In addition to catching exceptions, you can simply throw them. Throwing an exception simply sends the exception back to the function that called the current function, so that the previous function can handle the exception. The following example illustrates this.

Example 7.3

Step 1: Enter the following code in your favorite text editor.

```
#include <iostream>
using namespace std;
// function prototypes.
float divide_number(float , float);
int main()
{
 float dividend,divisor,answer;
 try
    {
```

```
cout << "Please enter a number \n";
cin >> dividend;
cout << "Please enter a number \n";
cin >> divisor;
answer = divide_number(dividend,divisor);
cout << dividend << " divided by ";
    cout << divisor << " is " << answer;
     }// end of try
     catch(...)
     {
     cout << "oops, there is an error!";
     }// end of catch
 return 0;
}
float divide_number(float num1, float num2)
{
 try
 {
   float answer;
   answer = num1/num2;
   return answer;
 }
 catch(...)
 {
     throw;
 }//end of catch
```

Step 2: Compile the code.

You can see that exceptions are not handled in the various functions, only in the main function. Some programmers prefer to centralize their error handling in this way.

SAVING ERRORS TO LOGS

Sometimes it is simply not helpful to display error messages to the user. If the error is not critical, it's more likely to simply confuse your end user without accomplishing anything useful. In fact, it is not prudent to display all messages to the user. One answer to this is to simply log the error messages to a flat file. Later, support personnel can read the log and see what has been occurring and perhaps diagnose what is wrong. You can even combine a message to the user with a log entry if you wish. The following example shows you how to log error messages to a file.

Example 7.4

Step 1: Enter the following code into your favorite text editor.

```cpp
#include <fstream>
#include <iostream>
using namespace std;
void logerror(int);
int main()
{
 int answer, divisor, dividend;
 try
 {
   cout << "Please enter an integer \n";
   cin >>divisor;
       cout << "Please enter another integer \n";
       cin >> dividend;
       if(dividend ==0)
     throw 0;

       answer = divisor/dividend;
     cout << answer;
       return 0;
 }
 catch (int i)
 {
   cout << "You cannot divide by zero";
       logerror(0);
 }
 catch (...)
 {
   logerror(1);
 }
}
void logerror(int type)
{
fstream myfile;
myfile.open("error.txt",ios::app);
if (type==0)
        myfile<<"Division by zero error \n";
else
        myfile<< "INdeterminate error \n";
return;
}
```

Step 2: Compile the code.

Step 3: Run the code. You should see something like what is depicted in Figure 7.3.

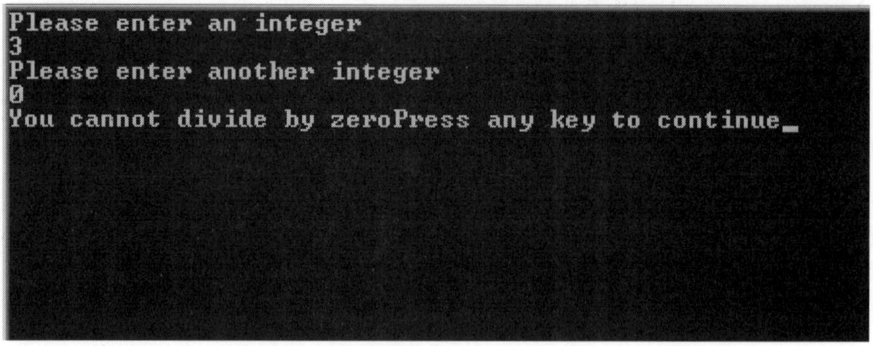

FIGURE 7.3   Writing errors to a log.

Now, if you open the file called `error.txt` on your hard drive, you should see something similar to Figure 7.4.

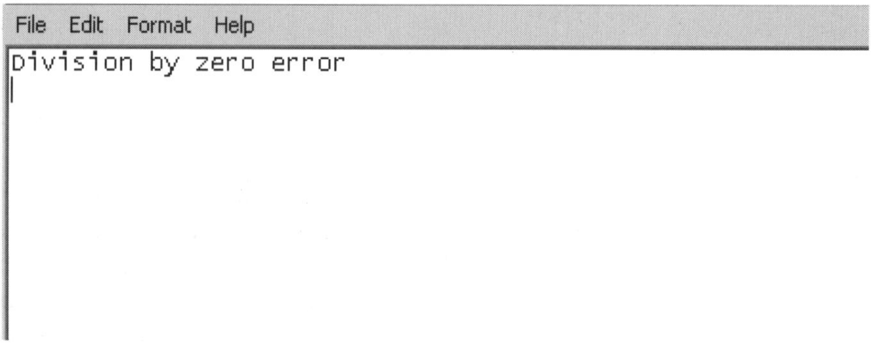

FIGURE 7.4   Viewing an error log.

This code sample has illustrated a number of important things. The first, and most obvious, is that you can log errors to a flat file if you so desire. But what you have also seen in this example is a variety of techniques from the past several chapters brought together. If you view each chapter and lesson as a separate and discrete entity, then you will miss the real point of this book. You should work to combine different techniques so that they work together to create useful applications.

GENERAL TESTING AND DEBUGGING

As this chapter has already pointed out, you will want to use extensive error handling to take care of any errors that occur during the execution of your code. However, an even better idea would be to properly test your code before you send it out. Good testing is a vital skill, and one that is rarely discussed in introductory C++ books.

There are three types of data you should test, and there are three types of testing you should do. The three types of data are: valid data, invalid data, and extreme data. The first, valid data, is simple. Type in the sort of data that your program is expecting and see whether it works. That is something most programmers, even beginners, address. The second area, invalid data, is sometimes overlooked. What happens if your program asks for an integer yet the user types in "A"? And, finally, you should check what happens with extreme data. If you ask the user to enter their age, and they enter 300 instead of 30, does your program just blindly go along with that clearly fallacious input? Or is your program smart enough to warn the user that their input is simply not possible? The discussion of data validation, later in this chapter, will show you how to accomplish this. For now, remember, when testing your program, to make certain that you test with all three types of data.

Now that you know what type of data to test, we can discuss how to test. There are three types of testing: unit testing, stress testing, and beta testing. Unit testing literally means to test the software on a single unit, normally your own PC. You would test all three types of data, but you would simply test them on your own machine. Stress testing involves pushing the program to its limits. If it is a multiuser program, then put a large number of users on it simultaneously. If its designed to handle 50 calculations per hour, then push it to at least 50 (actually going 10% beyond 50 would be even better). And, finally, we have beta testing. This involves simply getting your program to a handful of users and getting their feedback.

A lot of software today is inadequately tested. The news is filled with reports of software flaws, some causing significant problems. It is absolutely vital that you thoroughly test your software.

As you test your software you may find flaws. These are usually referred to in the industry as "bugs." It is amazing how many programmers are not very skilled at locating such bugs. Most programmers can easily fix a problem once it is found, but have difficulty finding it. There are some simple techniques you can use to find bugs in your programs.

One technique is to selectively comment your code. If the compiler tells you that there is an error, then simply comment out half your code and see if you still get the error. If you do still receive the error, add one half of the remaining code to your commented section and try to compile again. If you do not still receive the error, remove the comments from the one half you already commented. The following code has an error and generates a compiler error message syntax error : missing ; before identifier cout. Obviously, there is a problem with a missing semicolon before a cout statement, but which one? You could reread them all, and in a small program, that's probably the best thing to do. In a large program, however, doing this will not work. Let's take the section that has cout statements (the main function) and comment out half of it and see what happens.

```cpp
// Include statements
#include <iostream>
using namespace std;
float cube_number(float num);
float square_number(float num);
int main()
{
  float number;
  float number3;
  cout << "Please enter a number \n";
  cin >> number;
  if (number > 0 && number < 100)
  {
number3 = cube_number(number);
cout << number << "cubed is "<< number3;
  }// end of if
  else
  {
number3 = square_number(number)
cout << number << "square is "<< number3;
  }// end of else

  if (number3 <10000 || number3 ==2)
  {
```

```
 number3 = square_number(number3);
 cout << number3 << "squared is "<< number3;
  }// end of second if
 return 0;
}// end of main
float square_number(float num)
{
 float answer;
 answer = num * num;
 return answer;
}
float cube_number(float num)
{
 float answer;
 answer = num * num * num;
 return answer;
}
```

To find that error, comment out half the code and then try to compile
again.

```
int main()
{
  float number;
  float number3;
  cout << "Please enter a number \n";
  cin >> number;

  if (number > 0 && number < 100)
  {
number3 = cube_number(number);
cout << number << "cubed is "<< number3;
  }// end of if
/* BEGIN COMMENTING OUT CODE FOR TESTING
  else
  {
number3 = square_number(number)
cout << number << "square is "<< number3;
   }// end of else
   if (number3 <10000 || number3 ==2)
   {
number3 = square_number(number3);
cout << number3 << "squared is "<< number3;
   }// end of second if
```

```
   END COMMENTING OUT CODE FOR TESTING */
  return 1;
}// end of main
```

After compiling the code again, no errors are found. The problem, therefore, is in the section that was commented out. This technique can help you to drastically shrink the area you have to examine for a flaw. When writing larger programs, this will be helpful.

It is also important to look at the area surrounding an error. If you use one of the commercial IDEs such as Microsoft *Visual* or Borland *C++ Builder*, then the IDE can let you go directly to where the error is generated. However, sometimes the error is actually caused by code prior to where the error shows up. For example, a missing semicolon at the end of one line will cause the compiler to try to execute the next line as if it were part of that line. Thus, a good rule of thumb is to check the preceding four or five lines of code. Also remember the most common errors are the following.

1. Spelling
2. Missing semicolons
3. Forgetting to close a bracket around a block of code
4. Forgetting that C++ is case-sensitive

Always look for these errors. In fact, whenever you finish writing a significant section of code, it's a good idea to give it a quick read-through to search for these types of errors.

Another commonly used technique is to use cout statements at various locations in your code to print-out the value of certain variables. If you have a calculation that is not performing correctly, then this may be your best debugging technique. If you can see the values of the pertinent variables at each stage in the process you will be able to identify precisely where the problem occurs.

These are just a few "tricks of the trade" that might help you in debugging your applications. The important thing to remember about these techniques, and others, is that they are for debugging, not producing code. When you are finished debugging and testing, make sure you uncomment any code blocks that you commented out, and remove any cout statements that you put in for debugging purposes.

SUMMARY

This chapter has explored a valuable cornerstone of C++ programming, exceptions. Errors are going to occur, if you do not have some error handling in your program then your program will crash when those errors do occur. The chapter also discussed data validation and error handling, as well as testing. These fundamental skills are vital for all programmers, so please take the time to master them.

REVIEW QUESTIONS

1. What is the syntax in C++ to trap all errors?
2. What are the three general types of errors?
3. What are the four most common mistakes in C++?
4. List three of the basic exception classes in C++.
5. What does it mean to throw an exception?
6. What happens if you forget to terminate a statement with a semicolon?
7. What is the purpose of the catch block?
8. What is an exception?
9. What happens if an exception occurs that is NOT handled by a try-catch block?
10. What type of code can you place between the end of the try and the beginning of the catch?

User-Defined Data Types

IN THIS CHAPTER
• • • • • • • • • • • • • •

- Structures
- Typedefs
- Unions
- Enumerations
- Bit Fields

Throughout this book you have been using a variety of different variables of different types (int, float, bool, etc.). However, all these data types have something in common; they each store a single value. That statement may seem so obvious as to not need stating. However, having a variable store only a single value is not the only option you have when programming with C++. Another option is to create user-defined data types. User-defined data types are data types that you create. They allow you to store several, hopefully related, values under a single variable name. You literally create your own new data type. The difference between your data types and the ones that are part of the C++ language is that yours are complex and hold more than one value.

A structure is one of the most common sorts of user-defined data type. That means that you, the programmer, essentially create your own new data type to suit your needs. Structures are just one category of user-defined data types, but they are perhaps the most commonly used. Other user-defined types include typedefs, unions, and enumerations. Each of these user-defined data types will be explored and demonstrated in this chapter. The different categories of user-defined types are summarized in Table 8.1. Each will be explained in detail later in the chapter.

TABLE 8.1   User-Defined Types

User-Defined Type	Purpose	Example

Structure	A compound data type that simply combines several related pieces of data.	Struct employee
		{
		   char lastname[30];
		   float salary;
		};

Typedef	Provides a different name, or an alias for an existing data type.	Typedef int RETURN; Typedef double ANSWER;

Union	A combination of different data types.	Union NumericType
		{
		int iValue;
		long lValue;
		double dValue;
		}

Enumeration	An enumeration is a user-defined type that defines named constants.	enum test{value1,value2,value3};

Bit Field	A type of structure that allows you access to individual bits.	Struct status
		{
		   unsigned :cleartosend;
		   unsigned :ringing;
		}

Table 8.1 shows you the basics of the various user-defined data types. The following sections of this chapter will explore each of these in detail, with examples. The structure is perhaps the most commonly encountered user-defined data type, and it is essential to master it. However, the other user-defined types are also important and should not be ignored.

STRUCTURES
• • • • • • • • • •

A structure is really just a compound data type. It's a way of grouping several related variables under a common variable name. For example, if you were writing a program that involved employee data, you would have several individual pieces of data to store regarding a given employee. You would have a last name, first name, salary, and so on. If you place all that data into separate variables, it may become difficult to keep track of everything as your program grows in complexity. Which variables in your program are relevant to employee data and which are related to something else? This will become an even worse problem when you consider that other parts of your program might need similar data. For example, you would also need to store customer data, which would also have fields such as last name. Which last name goes with employees and which with customers? A structure provides you a fairly easy way to group all the data together. The following example illustrates a generic definition of a structure, and a specific one.

Generic example:

```
struct structname
{
  vartype varone;
  vartype vartwo;
};
```

Specific example:

```
struct employee
{
 char lastname[30];
 char firstname [30];
 float salary;
};
```

Once you have defined a structure you can then declare a variable of that structure type anywhere you wish. You create an instance of this data type just like you would create an instance of one of C++'s standard data types. You have the data type name followed by some name you wish to give your variable. Just like all variables, you have allocated a space in memory set aside to hold data of a certain type. The space you allocated depends on the size of your structure. You determine the size of a structure by totaling all of the individual data types. For example, if your structure contains two `int`s and one `char`, then it will be 2 * (four bytes per `int` + 1 byte for the character = 9 bytes).

```
employee myemployee;
```

You can then access the elements of the structure by using the name of the structure variable you created followed by a period and then the name of the element of the structure you wish to access.

```
myemployee.salary = 40000.44
```

As you can see, the structure definition is rather simple. The keyword `struct` is followed with the name of the structure. That name can be any valid C++ name, just as you would name any other variable. You then have brackets that define the boundaries of the structure. Because structures are often used in several places in a program, they are often defined in a header file. You can then simply include that header file in any source file where you wish to use the structure. The following example uses a structure to store data.

Example 8.1

Step 1: Type the following code into your favorite text editor.

```
#include <fstream>
#include <stdexcept>
#include <iostream>
using namespace std;
struct division
{
 float dividend;
 float divisor;
 float answer;
};
int main ()
{
 division localdivide;
 try
 {
   cout << "Please enter a number. \n";
   cin >> localdivide.dividend;
   cout << "Please enter another number.\n";
   cin >> localdivide.divisor;
```

```
    localdivide.answer =
localdivide.dividend/localdivide.divisor;

    cout << localdivide.dividend << "divided by " ;
    cout << localdivide.divisor << " is " <<
localdivide.answer;

}// end of the try blcok

catch(...)
{
  cout << "an error occurred!";
}
return 0;
    }
```

Step 2: Compile the code.
Step 3: Run the code. You should see something similar to Figure 8.1.

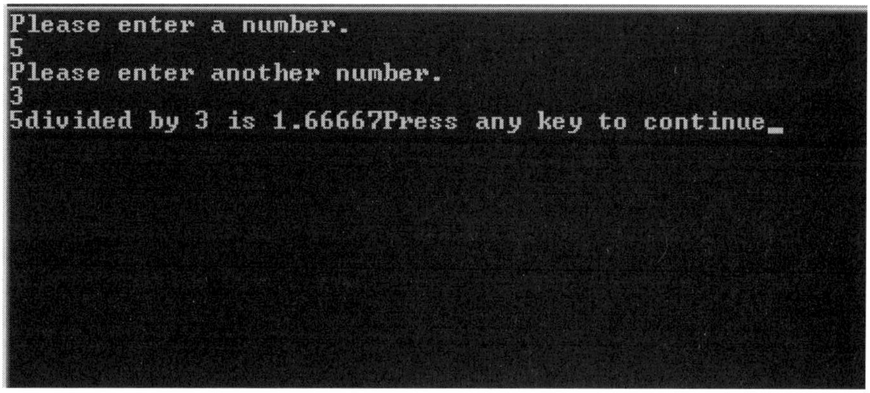

FIGURE 8.1 Using a structure.

As you can see, structures are not more difficult than standard C++ data types. However, a structure is a powerful tool because it allows you to group logically related data under a single name.

What is even more interesting is that a structure can also be passed into a function. This is of particular use when you have several variables that need to be passed into a function. If you consider our previous examples, you should recall a few that had several functions that required several items to be passed to them. Passing more than three or four variables to a function can be convoluted and confusing. It becomes an exceedingly inelegant-looking program. However, you can pass a single variable, a structure that contains all the various pieces of data you require. Passing them in a structure can be very convenient.

As was previously stated, you can use a structure just like any other data type. This means that not only can it be passed to a function as an argument, but also it can be returned. This can be very useful. Recall that functions can only return a single item. What if your function needs to give back several pieces of data? Returning a structure will allow that to happen. The following example illustrates this concept.

Example 8.2

Step 1: Place the following code in your favorite text editor.

```
#include <cmath>
#include <iostream>
using namespace std;
// define the structures
struct geometry_data
{
 float radius;
 double angle;
};
struct geometry_answers
{
 float area;
 double sine;
 double cosine;
 double tangent;
};
// prototype functions
geometry_answers compute(struct geometry_data mystruct);
int main ()
{
 geometry_data input;
 geometry_answers output;
 cout << "Enter the radius of the circle \n";
 cin >> input.radius;
 cout << "Enter the angle in rads \n";
```

```
 cin >> input.angle;
 output = compute(input);
 cout << " The area is "<< output.area << "\n";
    cout << " The sine of the angle is " << output.sine <<
"\n";

 cout << " The cosine of the angle is " <<
output.cosine << "\n";
 cout << " The tangent of the angle is " <<
   output.tangent << "\n";
return 0;
}
geometry_answers compute(struct geometry_data mystruct)
{
 geometry_answers answer;
 answer.area = 3.14f * pow(mystruct.radius,2);
 answer.sine = sin(mystruct.angle);
 answer.cosine = cos(mystruct.angle);
 answer.tangent = tan(mystruct.angle);
 return answer;
};
```

Step 2: Compile the code.

Step 3: Execute the code. You should see something much like what is shown in Figure 8.2.

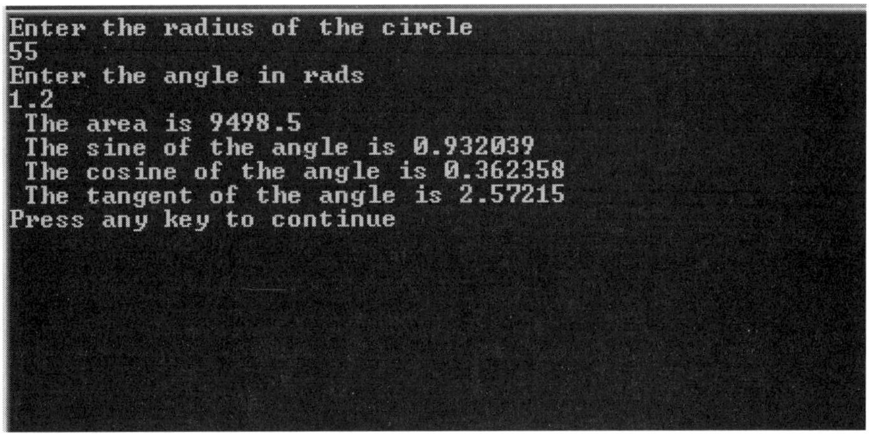

FIGURE 8.2   Structures as return types.

You can see how one structure is used to pass data into the function, then another structure is used to pass out the various answers. In this way, you can do several related geometric operations inside a single function and return all the relevant answers.

You can also pass that structure by reference (recall passing by reference was introduced in the final section of Chapter 3), then it becomes even more convenient. Let's look at an example.

Example 8.3

Step 1: Enter the following code in your favorite text editor.

```
#include <iostream>
using namespace std;
struct triangle
{
 float base;
 float height;
 float area;
};
void trianglearea(struct triangle &area);
int main ()
{
 triangle mytriangle;
 cout << "Please enter the base of a triangle \n";
 cin >> mytriangle.base;
 cout << "Please enter the height of a triangle \n";
 cin >> mytriangle.height;

 trianglearea(mytriangle);
 cout << "The area is " << mytriangle.area << " \n";

return 0;
}
void trianglearea(triangle &area)
{
 area.area = .5f * (area.base * area.height);
}
```

Step 2: Compile the code.
Step 3: Run the code. You should see something like Figure 8.3.

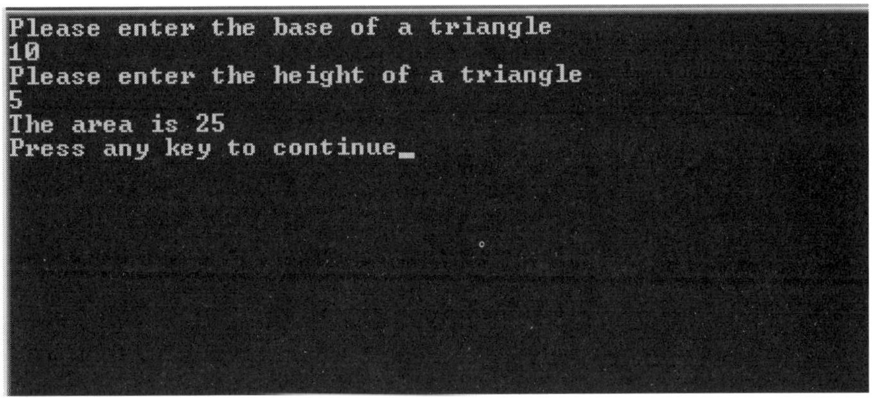

FIGURE 8.3   Passing a structure.

Thus, you can see that passing the entire structure by reference accomplishes two goals. First of all, you can pass multiple values in a single variable name. Second, you can get multiple return values stored in a single structure. Without the use of a structure, any function can only return one value. Even with a structure the function can still only return one variable, but now that variable can hold more values.

In an earlier chapter you were introduced to the use of functions such as memcpy in your applications; however, the use of these functions with structures is slightly different. As you have seen, a structure is a complex data type that holds several other simple data types. This has an effect when using some of the built-in functions in C++, such as memcpy. When you use memcpy with an array you simply pass it the destination array, source array, and size of array. When you use memcpy with a structure you must pass it the address of the structure. This is illustrated in the following example.

Example 8.4

Step 1: Enter the following code into a text editor and save it as *08-04.cpp*.

```
#include <memory>
#include <iostream>
using namespace std;
struct mystruct
{
```

```
 int i;
 int x;
 int y;
};

int main()
{
 mystruct source,destination;
 cout << "Please enter a number \n";
 cin >> source.i;
 cout << "Please enter another number \n";
 cin >> source.x;
 cout << "Please enter another number \n";
 cin >> source.y;
 memcpy(&destination,&source,sizeof(source));
 cout << destination.i << endl;
 cout << destination.x << endl;
 cout << destination.y << endl;
 return 0;
    }
```

Step 2: Compile and run your code. You should see something similar to Figure 8.4.

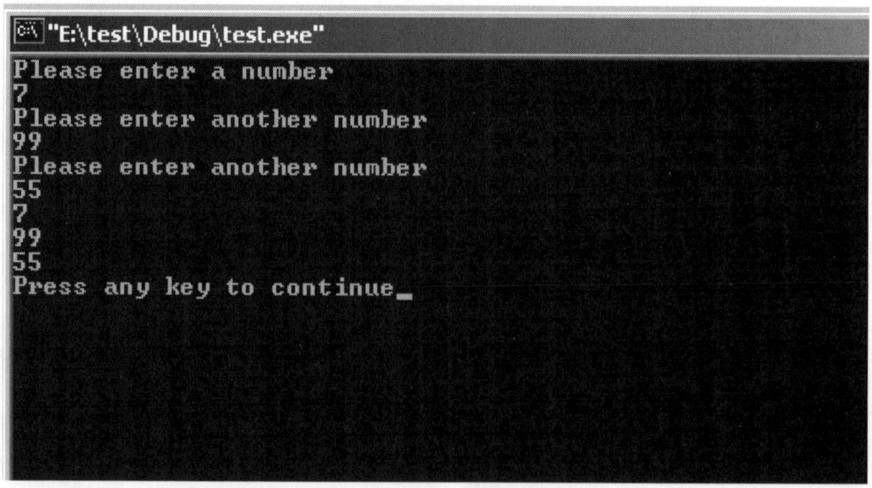

FIGURE 8.4   Structures and memcpy.

You will notice that we copied the contents from the source structure to the destination structure and then the contents of the destination structure are printed. This illustrates that the copy worked.

TYPEDEFS
• • • • • • • •

A structure is only one way that you can define your own types of variables in C++, the typedef is another. A typedef in C++ is a type of variable declaration. The purpose of the typedef is to create new types from existing data types. Because a typedef is simply a declaration, it can be intermingled with standard variable declarations. However, it is common practice to declare all typedefs before declaring the standard variables. It is also common practice to capitalize the first letter of a typedef to distinguish it from the built-in types, which all have lower-case names. Typedefs are often used to create synonyms for built-in types.

```
typedef int Integer; //The word 'Integer' can now be
                     // used in place of int
int k,j     // this creates two ints
Integer m,n; // this also creates two ints
```

In this example, a typedef was used to substitute for a built-in type. This is generally not recommended. Why not just use the built-in type? However, because it is used in code you may see, it is important for you to see this application of typedef. Usually, a typedef associates a type name with a more complicated type specification. Typedefs are often used with arrays. A typedef should always be used in situations where the same type definition is used more than once for the same purpose.

```
typedef int myarray[10]; //10 integers
myarray a,b; // This creates two arrays
int x[10], y[10]; //This does the same thing without
                  //typedef
```

Basically, the typedef is a way for you to create your own new labels for data types. You can use it with any data type, even as a synonym for an existing basic type such as int. It is recommended that you only use it with more complex data types such as an array. Typedefs can even be used for other types you create such as structures. Consider the following code sample.

Example 8.5

Step 1: Type the following code into your favorite text editor.

```cpp
#include <cmath>
#include <iostream>
using namespace std;
typedef double ANSWER;
// function prototypes
ANSWER squareroot(double number);
ANSWER cuberoot(double number);
ANSWER fourthroot(double number);
int main()
{
    int menu;
    double num;
    ANSWER response;
cout << "1. Square root \n";
    cout << "2. Cube root \n";
    cout << "3. Fourth root \n";
    cout << "Enter your selection\n";
    cin >> menu;
    switch (menu)
    {
    case 1:
    cout << "Enter your number \n";
     cin >> num;
    response = squareroot(num);
     break;
    case 2:
    cout << "Enter your number \n";
     cin >> num;
    response = cuberoot(num);
     break;
    case 3:
    cout << "Enter your number \n";
     cin >> num;
    response = fourthroot(num);
     break;
    default:
    cout << "Invalid entry\n";
```

```
        }
        cout << "The answer is " << response;
        return 0;
}
ANSWER squareroot(double number)
{
 return sqrt(number);
}
ANSWER cuberoot(double number)
{
 return pow(number,.3333);
}
ANSWER fourthroot(double number)
{
 return pow (number,.25);
}
```

Step 2: Compile that code.

As you can see, the typdef ANSWER can now be used in any math function that needs to return an answer. This can be useful for you when you have a particular data type that is used in a specialized way in a variety of places. You can also move the typedef to a header file so that you can include it in any program that you wish to use it in.

You should also notice a few other things about this particular example. In addition to illustrating the use of a typedef, it also shows you how to use the pow function in a creative way to get higher roots of numbers. For example, the fourth root of a number is the same as raising that number to the *th* power, or the .25 power. Although it is not the purpose of this book to teach you math, it is important that you learn to take various principles, combine them, and apply them in creative ways. That type of creative thinking marks the difference between mediocre programmers and great programmers.

UNIONS

So far we have looked at structures and typedefs as examples of user-defined types. There are still more user-defined types to examine. The one we will examine next is the union. A union allows you to handle an area of memory that could contain different types of variables. The syntax for unions is identical to that for structures. You can even contain unions within structures, or vice versa. The following is an example of a union data type.

```
union NumericType    // Declare a union that can hold
                     // the following:
{
intiValue;        // int value
long    lValue;   // long value
double dValue;    // double value
};
```

If you think about it, the reason these data types are called unions is that they are the union of several different data types. The following is an example using a union.

Example 8.6

Step 1: Enter the following code into your favorite text editor.

```
#include <iostream>
using namespace std;
union NumericType    // Declare a union that can
                     // hold the following:
{
    intiValue;        // int value
    long    lValue;   // long value
    double dValue;    // double value
};
int main()
{
 NumericType values;
 cout << "Please enter a number \n";
 cin >> values.dValue;
 cout << "Please enter another number \n";
 cin >> values.lValue;
 cout << values.dValue;
 cout << " \n";
 cout << values.lValue;
    cout << "press any key to continue\n";

    return 0;
}
```

Step 2: Compile your code.
Step 3: Run your code. You should see something similar to Figure 8.5.

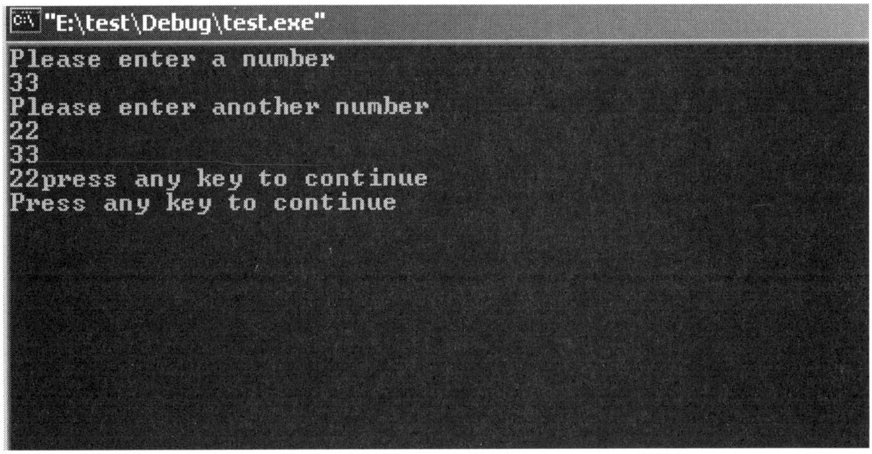

FIGURE 8.5 Using a union.

ENUMERATIONS

Now you have seen structures, typedefs, and unions. The next category of user-defined data type is the *enumeration*. An enumeration is a user-defined type that defines named constants. Enumerations are declared using the enum keyword. An enumeration is really a set of named integer constants that specify all the values that variables of this type may have. Note that these are named integer constants. That means that we give names to integer values.

The enumerated type is declared using the following form.

```
enum typename {enumeration list};
```

enum is a C++ keyword. The compiler sees this word and knows that the definition for an enumeration is coming up. The word typename can be any valid C++ name identifier. The enumerated list is a comma-delimited list of the items in the enumeration, enclosed in brackets, and terminated by a semicolon. The following is a generic example.

```
enum test{value1,value2,value3};
```

BIT FIELDS

Many of the user-defined data types we have examined so far are available, in one form or another, in other programming languages such as *Java* and *Visual Basic.* However, C++ offers one intriguing user-defined type that is not present in most languages. That type is the bit field. Normally, in any programming language, any variable is of some particular data type (`int`, `float`, `bool`, etc.) and takes up a certain number of bytes (remember that there are eight bits in a byte). However, C++ allows you to access individual bits. This is of particular importance when writing software that communicates with hardware, and in the telecommunications industry.

When you create a bit field, you create it like you would any normal structure. However, each individual element of the structure is declared as an unsigned, then given a name and a colon. After the colon is a number representing how many bits this particular element occupies.

It is beyond the scope of this book to explore either hardware programming or telecommunications programming. However, both of these areas of software development make extensive use of the C++ programming language, and of features such as bit fields. The following example should help illustrate the basics of using a bit field. Although the example is about checking the status of a serial line and taking appropriate action, the specific code for the particular actions to take is beyond the scope of this book and, therefore, the functions called are empty, with just comments in them. The studious reader can easily find appropriate serial line code examples on the Internet and expand this example. Our interest now is simply to study the bit field itself, not its potential applications.

Example 8.7

Step 1: Enter the following code into your favorite text editor.

```
#include <iostream>
using namespace std;
// declare a bit field
struct status
{
    unsigned changeinline: 1;
    unsigned cleartosend:1;
```

```
        unsigned inactive:1;
        unsigned ringing:1;
        unsigned signalreceived:1;
};
// function prototypes
void dialnumber();
void answerphone();
void senddata();
int main()
    {
    // delcare an instance
    // of the bit-field
status linestatus;
    if (linestatus.cleartosend)
    senddata();
    if (linestatus.inactive)
    dialnumber();

    if(linestatus.ringing)
    answerphone();
    return 0;
    }
void dialnumber()
{
// place code here to dial
}
void answerphone()
{
 // place code here to answer the phone
}
void senddata()
{
 // place code here to send the data
}
```

Step 2: Compile that code.

The important thing to realize is that the bit field allows you to directly speak the same language as PC hardware, and telecommunications lines. These devices do not send data in types such as int, long, or double. They send a stream of individual bits. C++ allows you to directly work with those individual bits in a way that few other languages do.

SUMMARY

In this chapter you were introduced to a wide variety of user-defined data types. Specifically we looked at structures, enumerations (also referred to simply as enums), unions, and bit fields. By far the most attention was given to structures. You were shown how to create your own data types to use in your programs. Of these, the structure is perhaps the most commonly encountered and the most important. Make certain that you are at least basically familiar with all these data types, but especially the structure. When you begin Chapter 9, the structure is expanded to form classes for object-oriented programming.

REVIEW QUESTIONS

1. What is the purpose of a structure?
2. Can a structure contain different data types?
3. What is the reason for using a typedef?
4. Where are structures usually defined?
5. How are instances of your structure made?
6. Can structures be passed as arguments to functions?
7. A typedef is a type of_____.
8. Can a structure by used as a return type?
9. What advantage is there to passing a structure by reference?

Pointers

IN THIS CHAPTER
• • • • • • • • • • • •

- Pointer Basics
- Pointer Operations
- Pointers to Pointers
- Pointers to Functions
- Pointers to Structures
- Initializing Pointers

Pointers are a powerful, but troublesome part of C and C++. Most beginning C/C++ programmers have significant trouble understanding and appropriately using pointers, so please pay close attention to this chapter. The first, and most obvious question, is what, indeed, is a pointer? This question is probably foremost in your mind at this particular point in time. A *pointer* is a special type of variable, one that literally points to an address in memory of another variable. If you think about standard variables, they are labels used to identify small segments of memory set aside to hold data of a particular type. A *pointer* simply points to the segment of memory occupied by another variable. Whereas a standard variable holds a value, the value stored at an address in memory, a *pointer* simply holds the address of another variable. You probably noticed that this paragraph provided three different but similar definitions for the pointer. All three of these definitions say the same

thing, but in slightly different ways. It is hoped that one of these definitions will strike a chord with you and you will comprehend the concept.

POINTER BASICS

As previously stated, a variable is a small section in memory set aside to
hold data of a specific type. When you write:

```
int j;
```

you have just set aside 4 bytes of memory and you are referring to that 4
bytes by the variable j. The variable j represents whatever value is currently stored in those 4 bytes of memory.

Declaring a pointer is almost the same. You still declare a data type,
just as you do with standard variables; however, you must add the
asterisk (*) symbol to denote that it is a pointer. You can place the asterisk
immediately after the data type, or before the variable name. Either
method is a valid way of declaring a pointer variable. For example, both
of the following are valid pointer declarations.

```
int* k;
int *k;
```

The asterisk(*) is called a reference operator. This name makes perfect
sense, because it, indeed, is a reference to an address in memory. The *
operator can be literally translated to "the address pointed at by," The
way you use a pointer is that you have it set equal to the address of
another variable. Recall from Chapter 1 that the ampersand (&) operator
denotes an address in memory, or is the "address of" operator. Let's look
at the following examples.

```
int j;
int *k:
k = &j;
```

Now the pointer k represents the address of j. More specifically it is
pointing to the first byte of the 4 bytes that j occupies. This is illustrated
in Figure 9.1.

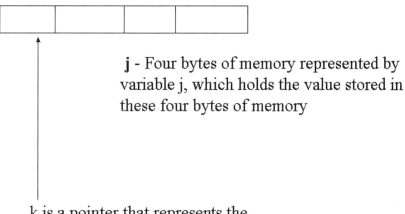

j - Four bytes of memory represented by variable j, which holds the value stored in these four bytes of memory

k is a pointer that represents the address of the first byte.

FIGURE 9.1   Pointers.

This is important to remember for several reasons. The first being that if you print-out j you will get a numeric value. However, if you print-out k you will get a rather long number that represents an address in memory. The following example illustrates this.

Example 9.1

Step 1: Write the following code in your favorite text editor.

```
#include <iostream>
using namespace std;
   int main()
   {
     int i;
     int* j;
     j = &i;
     i = 4;
     cout << "i is " << i;
     cout << "\n j is " << j << "\n";
return 1;
   }
```

Step 2: Compile the code.

Step 3: Run the code. You should see something like what is shown in Figure 9.2

```
i is 4
 j is 0x0012FF7C
Press any key to continue_
```

FIGURE 9.2   Printing pointers.

You will note that when the pointer variable is printed out, a rather long number is displayed on the screen. This is NOT the value stored at a particular address in memory; rather, it is the actual numerical address of the first byte of the variable that the pointer points to. In addition to illustrating that pointers represent addresses, this should also help you understand why we use variables. Which is easier to remember, the variable j, or some lengthy address in memory?

HINT!
. . . .

Pointers are typed. This means you must tell the compiler what type of variable your pointer is pointing to.

The fact that pointers are typed means that the following code would generate an error.

```
char c = '0';
   int *p = &c;
```

This code will generate an error because you cannot use an int pointer to point to a variable of type char. Remember that you must use a pointer of the same data type as the variable it is pointing to.

POINTER OPERATIONS

Because a pointer is actually an address in memory, rather than a value, it should make sense to you that normal operations on a pointer behave quite differently. For example, with a normal variable if you use the increment operator on it (++), you increase its value by one. With a pointer, you actually move the pointer over a number of bytes equal to the size of the data type. Because an integer is 4 bytes (at least on 32-bit operating systems), if you have an integer pointer and do the increment operation, then it will move the pointer over 4 bytes in memory. So, for example, if you move an `int` pointer over 4 bytes, it will no longer be pointing at that same `int`. It will have moved past the 4 bytes occupied by the `int` variable and will be pointing at some space in memory, the contents of which are not known.

WATCHOUT!

 If you move your pointer to some unknown memory location and then do some operation on it, you might be interfering with the operations of some other program on your PC, including the operating system.

To conduct arithmetical operations on pointers is somewhat different than to conduct operations on other data types. To begin with, you can only use addition (+), subtraction (-), increment (++), and decrement (–) operators. You cannot use multiplication or division on a pointer. Because a pointer contains an address in memory, multiplication and division do not make sense. A pointer holds an address, if you multiply that by 4, you get a number that may represent more memory that your computer even has. But both addition and subtraction have a different behavior with pointers then they do with normal data types. The addition operator causes the pointer to move a certain number of bytes to the right. The subtraction operator causes the pointer to move a certain number of bytes to the left. The number of bytes moved is the number used in the expression times the size of the data type. Consider the following code.

```
int a;
int *p;
```

```
p = &a;
p =p+2;
```

What actually happens in the fourth line is that the pointer moves over bytes. It is not adding 2 to the value of a. It is, instead, pointing to the last 2 bytes of the integer a. This is illustrated in Figure 9.3.

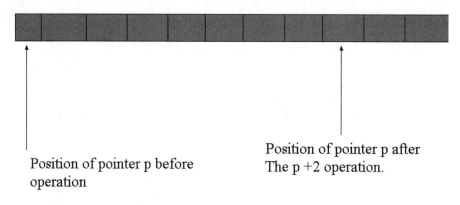

Position of pointer p before operation

Position of pointer p after The p +2 operation.

FIGURE 9.3   Adding to a pointer.

You can see that what is occurring is that the pointer is simply moving over. Now this may not seem to be of much use with standard variables, but it is useful with arrays. You will frequently see pointers used in conjunction with arrays. You should recall arrays from Chapter 1. (If you need a refresher, then please return to that chapter and reread the section on arrays.) If you have a pointer with an array, it starts off pointing to the first element of the array. Any arithmetic operations you perform on it will simply move the pointer forward and backward through the array. The following example should help you understand this.

Example 9.2

Step 1: Enter the following code in your favorite text editor.

```
#include <iostream>
using namespace std;
int main ()
{
```

```
int i[5];
int *p;
p = i;
for(int j = 0;j<5;j++)
{
    i[j] = j;
p++;
    cout << p;
    cout << "\n";
}
return 0;
    }
```

Step 2: Compile your code.

Step 3: Run the application. You should see an image much like Figure 9.4.

FIGURE 9.4   Pointers and arrays.

You will notice that what you see is a series of addresses in memory. The pointer points to those addresses, and by moving the pointer you are shifting it to the next element in the array. This is also an excellent time to remind you of the grave danger you can face using a pointer. A pointer will attempt to access whatever memory address you point it to . . . even if that goes beyond the bounds of your array, or even your application! Consider the following example.

Example 9.3

Step 1: Enter the following code in your favorite text editor.

```
#include <iostream>
using namespace std;
int main ()
{
 int i[5];
 int *p;
 p = i;
 for(int j = 0;j<5;j++)
 {
      i[j] = j;
 }
 for(int k = 0;k<7;k++)
 {
      p++;
      cout << p;
      cout << "\n";
 }
 return 0;
  }
```

Step 2: Compile the code.

This code does not generate an error. C++ allowed you to point the pointer beyond the bounds of the array. So what is the pointer now pointing at? There is no way to know. It might be pointing at another variable in your program, some part of another program running on your PC, or even at some part of the operating system. This is why pointers, if not used carefully, can be quite dangerous.

Remember that programs are not necessarily loaded into contiguous blocks of memory. Right next to your array, in memory, could be a variable from another program, or even from a subroutine of your operating system. Not paying attention to pointers is one way that a C++ programmer can get into a lot of trouble and cause some significant problems on a system. Another thing to remember when working with pointers is to always initialize them to some value. If you simply declare a pointer then try to use it without first pointing to some address in memory, then it will be pointing at an arbitrary address in memory. That could be very problematic. A common technique is to initialize the pointer so that it points at some null value.

```
int *x = 0;
```

POINTERS TO POINTERS

So far we have simply looked at pointers to standard variables. However, C++ also allows the use of pointers that point to other pointers. A pointer to a pointer is often referred to as multiple indirection. Because a pointer is simply an indirect method of accessing a variable, then a pointer to a pointer would be multiple indirection. To create a pointer to a pointer, we only need to add an asterisk (*) for each level of reference.

```
char x;
char * y;
char ** z;
x = 'a';
y = &x;
z = &y;
```

The truth is that you can program for years and never have a particularly compelling need to implement pointers to pointers. In addition, pointers to pointers will not be used in this book.

POINTERS TO FUNCTIONS

C++ also allows another type of pointer; this one is more useful than pointers to pointers. C++ allows you to create pointers to functions. The greater utility of this is for passing a function as a parameter to another function. That means you can pass an entire function as an argument for another function. To declare a pointer to a function we must declare it like the prototype of the function, but enclosing between parenthesis () the name of the function and then placing a pointer asterisk (*) before the name of the function.

```
// pointer to functions
#include <iostream>
using namespace std;
// function prototypes
int subtraction (int a, int b);
int addition (int a, int b);
//function pointer prototypes int (*minus)(int,int) =
subtraction;
int operation (int x, int y, int (*functocall)(int,int))
{
```

```
int i;
i = (*functocall)(x,y);
return (i);
}
int main ()
{
 int m,n;
 m = operation (5, 5, addition);
 n = operation (50, m, minus);
 cout <<n;
 return 0;
}
int addition (int a, int b)
{
    int answer;
    answer = a + b;
    return answer;
}
int subtraction (int a, int b)
{
    int answer;
    answer = a - b;
    return answer;
}
```

In the example, the word minus is used as a pointer to a function that has two parameters of type int. You can now pass that function's pointer to another function.

```
int (* minus)(int,int) = subtraction;
```

In this case, you pass the functions pointer to a function called operation, which will do whatever function is passed to it (be it addition or subtraction). This is a rather powerful feature that C++ provides you.

POINTERS TO STRUCTURES

As you have probably seen, we can use a pointer to any data type, or even to a function. If you think about this for just a moment it makes sense. Everything in your program is loaded into memory. A pointer is simply an indicator of an address in memory. Your variables are places in memory, other pointers are referencing a place in memory, and functions are loaded in memory. Because all this is at some location in memory, it makes sense that you should be able to have a pointer variable that references that location in memory.

Because a structure is a complex data type, you can also have a pointer to a structure variable. In fact, pointers to structures are a commonly used technique. Recall from Chapter 7, when we first discussed structures, that a structure variable allows you to pass several pieces of data, and allows the function to return several pieces of data. A pointer to a structure allows you to point to that space in memory, just as you would point to any other variable's space in memory.

There is one slight change when dealing with pointers to structures. Normally you access an element of a structure by giving the structure name, a (.), then the name of the element. For pointers to structures you must use a -> rather than a pointer. Let's examine an example that should clarify this.

Example 9.4

Step 1: Write the following code into your favorite text editor.

```
#include <iostream>
using namespace std;
struct account
{
   int accountnum;
   float balance;
   float interestrate;
};
int main()
{
  account myaccount;
  account *ptraccount;
  //set the balance of the structure
  myaccount.balance = 1000;
  //initialize the pointer
  ptraccount = &myaccount;
  //change the pointers balance
  ptraccount->balance = 2000;
  //print out the structures balance cout << myaccount.balance
<< "\n";
  return 0;
}
```

Step 2: Compile the code.

Step 3: Run the program. You should see something similar to Figure 9.5.

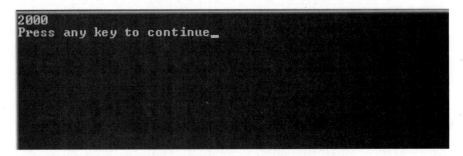

```
2000
Press any key to continue_
```

FIGURE 9.5   Pointers to structures.

It is important to note, in this example, that when you change an element of the pointer to the structure, you are actually changing the element of the structure itself. This is because, like all points, a structure pointer is simply redirecting any commands or operations to the address in memory it points to.

INITIALIZING POINTERS

There is one last, but critical issue to address regarding pointers. Before you use a pointer you must initialize it. You initialize a pointer in one of two ways. The most common way is to set the pointer equal to the address of some standard variable. With pointers to standard variable types you can also set the pointer equal to 0, as in the following example.

```
int *ptr = 0;
```

If you attempt to utilize a pointer that has not been initialized, by one method or the other, then you will generate an exception and probably crash your program. Because a pointer is a reference to an address in memory, if you don't initialize it then it is pointing to some indeterminate address in memory. The following short example illustrates what happens when you use a pointer that has not been initialized.

Example 9.5

Step 1: Type the following code into your favorite text editor.

```
#incl178ude <iostream>
using namespace std;
struct account
{
 int accountnum;
 float balance;
 float interestrate;
};
int main()
{
account myaccount;
account *ptraccount;
ptraccount->balance = 1000;
    return 0;
  }
```

Step 2: Compile that code.
Step 3: Run the code. You should see something like what is shown in Figure 9.6.

FIGURE 9.6   Using a pointer that is not initialized.

It is absolutely critical that you do not use a pointer that has not been initialized.

SUMMARY

This chapter introduced you to one of the most daunting topics in C++, pointers. Pointers have long been the bane of beginning programming students. Remember that a pointer is simply a variable that points to an address of another variable. Put another way, a pointer contains an address rather than a value. In this chapter, you saw how to use pointers, how not to use pointers, how to use pointers to other pointers, and how to use pointers to functions. We also discussed using various operators on pointers.

REVIEW QUESTIONS

1. What does it mean to say that pointers are type-specific?
2. What happens when you have an int pointer named p and you execute the expression p++?
3. What does a pointer point to?
4. Where is one place that pointers are often used?
5. Arithmetic operations on a pointer do what?
6. What is the & operator?
7. What is **p?
8. Why would you want to create a function pointer?
9. What is a pointer?
10. Why don't pointers support multiplication?

Object-Oriented Programming

This section will take you into the world of object-oriented programming. The main reason why C++ was created was to bring object-oriented technology to the world of C programming. Some beginning
programmers are intimidated by object-oriented programming but there
is no need to be. Each of the chapters in this section will carefully explain
this exciting area of C++ programming to you.

Classes

IN THIS CHAPTER
· · · · · · · · · · · · · ·

- Object-Oriented Theory
- What Is a Class?
- Creating and Using a Class
- Pointers to Classes
- Classes and Exception Handling
- Constructors and Destructors
- Arrays and Pointers with Classes

The entire purpose of C++ is to add object orientation to the C programming language. The first 9 chapters have illustrated simple, non-object-oriented C++. You will still use all these techniques in the rest of this book, but now object-oriented programming will be added. This chapter will introduce you to the concepts of object-oriented theory as well as to the basic unit of object-oriented programming: classes.

OBJECT-ORIENTED THEORY
· ·

The first question we should address is: What is object-oriented programming? The answer is simple: It is programming that is meant to more accurately represent objects in the real world, by using objects in code. Object-oriented programming is simply about accurately modeling the real world in your programs. It's also about code reusability. The heart of object-oriented programming is to create an object, in code, that has certain properties and methods. This is not unlike the real world. You car is an object. It has certain properties such as color, number of doors, and the like. It also has certain methods such as accelerate, brake, and so on. You use that object by executing one or more of its methods. An object in programming is exactly the same. It has certain properties that describe it, and certain methods you can use to do certain things with it.

There are four principle concepts that form the foundation of object-oriented programming. These concepts are abstraction, encapsulation, inheritance, and polymorphism.

Abstraction

Abstraction means you can deal with things in a general sort of way, and only get as specific as you need to at a given point in time. For example, we could discuss the general characteristics of an automobile without dealing with a specific make and model. Another, and more formal, way to define abstraction is that it is the act of specifying a general interface, hiding implementation details.

Encapsulation

Encapsulation is placing the data and the functions that work on that data in the same place. In procedural programming, like what you have seen in the first 8 chapters of this book, it is not always clear which functions work on which variables. The more complex the program, the more unclear this relationship becomes. With object-oriented programming you place the data and the relevant functions together in the same object. Thus, which functions work on which data is abundantly clear.

Inheritance

One of the most useful aspects of object-oriented programming is code reusability. One object can inherit the public functions and data from another object.

Polymorphism

Once you have inherited a function from another object, you can manipulate it and change it. Thus, one function can take many (poly) forms (morph) (i.e., polymorphism).

You should probably try to commit these four principle concepts and their explanations to memory. At this point, you are probably still a bit fuzzy on exact definitions of these terms. Don't let that worry you too much. Most people have a little difficulty when they first encounter object-oriented programming. Only through continued use and reexplanation of the principles involved, will you become comfortable with these concepts. In this chapter and the next few chapters, each of these concepts will be explored in more detail.

WHAT IS A CLASS?

As you have already seen, object-oriented programming represents real-world objects in programming structures. It does this by using classes. A *class* can be defined as a template for creating objects. A more practical description might be that it's a structure with methods. If you will recall structures from Chapter 7, you can think of a class as just a structure that also has methods. Once you have created a class, you can then declare an instance of it, just as you would create any other type of variable. Consider the following example.

Example 10.1

Step 1: Please enter the following code into your favorite text editor.

```
#include <iostream>
using namespace std;
class myfirstclass
{
public:
 int number;
 void greeting();
};
void myfirstclass::greeting()
{
```

```
for(int i = 0; i<= number;i++)
{
  cout<< "Hello World \n";
}


}
int main ()
{
    myfirstclass myfirstobject;
    myfirstobject.number =3;
myfirstobject.greeting();
return 0;
    }
```

Step 2: Compile the code.
Step 3: Run the code. You should see something similar to Figure 10.1.

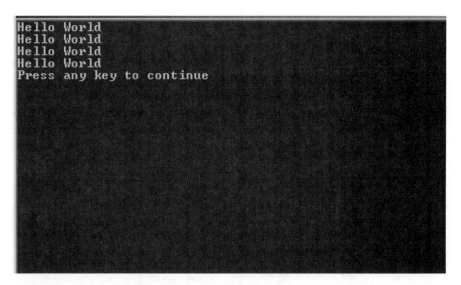

FIGURE 10.1   My first class.

There several things in this code that should be interesting you. Let's examine the code one piece at a time. To begin with you should note the overall framework of a class is not unlike a structure. You have the key-word class identifying that this is a class, then you have the name that we wish to give it. You then have brackets showing the boundaries of the class.

```
class myclass
{
};
```

So far there is really no difference between a structure and a class. As was pointed out earlier, a class is really just a structure with methods. The first difference is the word *public*: Anything, be it a function or a variable, listed before the word *public* will be considered private. (The difference between public and private will be discussed momentarily.) Next, we notice a few variables are declared. These variable declarations are not different from any other such declarations you have seen so far in this book. By default, everything in a class is private unless you specifically declare it as public or protected. Everything in a structure is public unless you specifically declare it as private or protected.

The real oddity is how functions are declared and associated with a class. You first must designate the functions return type, just as you would with any function. But then, before you give the name of the function, you must give the class name followed by two colons :: then the name of the function. The two colons are referred to as a scope operator. You may recall them from our brief discussion on namespaces. A scope operator simply designates that the item (be it a function, variable, etc.) immediately following it is associated with (or part of) the object preceding it.

When the code reaches the main function we can now declare variables of the type myfirstclass. We have essentially created a new data type; however, this data type not only stores data but also has functions. Then simply using the name of the variable followed by a . and the function or variable we wish to access, we can either set the variables values, or call the function.

You should notice that what we have now created is an object much like objects in the real world. This object has variables; in the world of object-oriented programming (OOP) we call these *properties*, and it has functions, which we refer to as *methods*. Most of the objects around you have properties and methods. As previously mentioned, a car has properties such as number of doors, color, model, and the like. It also has methods such as accelerate, brake, and so on. In object orientation we take the variables or data we wish to work on and place them in a class with the methods that will work with that data. This is the concept of encapsulation introduced earlier in this book.

The difference between private items and public items is simple. Public items can be called from outside the class, as you saw in our previous example. Private items cannot be called from outside the class. They are called from other methods of the class. Your car has a number of private methods in its engine. These are not readily available to the outside. When you initiate the public accelerate method, several private methods are then called to increase fuel flow and other items necessary to accelerate. Public or private determines how the particular member function or variable can be accessed from outside the class, they are therefore called access modifiers. There are actually different possibilities for the access modifier of a method or property of a class; these are shown in Table 10.1.

TABLE 10.1   Access Modifiers

Access Modifier	Purpose
Private	This means that the member variable or function cannot be accessed, or even viewed from outside the class.
Public	Public members can be accessed from outside the class.
Protected	Protected members cannot be accessed from outside the class, but they are inherited by subclasses.

CREATING AND USING A CLASS

A class has a level of complexity somewhat beyond that of structures. This means that although you can create a class inside a source file with the rest of your program, this can become quite cumbersome. For this reason, classes are often defined in header files. The combination of classes and header files makes for a high level of code reusability. A class can be created that has a particular set of functionality needed to accomplish certain tasks. With it being defined in a header file, you can then simply include that header file in any source file where you need that functionality, then declare an instance of that class. You will often see programmers using classes created by other programmers, and in some cases not even having a particularly good idea of how that class accomplishes its goal. In fact, you have already done that. Think about this for a second . . . do you know *how* cout takes your typed commands and displays them on the screen? Probably not.

Another important item to keep in mind is that a class is a template for creating objects. When you create a class you are essentially creating a new data type, one that is complex and has its own functions or methods. You cannot simply start using the class, you must first create an instance of it. This is not different from any other data type. You cannot, for example, write the following.

```
int = 7;
```

The int is a data type, and not an actual variable. You must first create a specific int before you can assign values to it.

```
int j = 7;
```

The same is true of your classes. The process of creating an instance of a class has several names, all of which mean the same thing. You will hear this referred to as instantiating a class, creating an instance of a class, or creating an object from a class. All of these simply mean that you are creating a specific variable of the type of your class.

Let's examine another example of a class; this one will probably be more useful than our first example.

Example 10.2

Step 1: Enter the following code into your favorite text editor and save it as *010-02.h.*

```
#include <cmath>
class myfirstclass
{
      float answer;
      double danswer;
public:
 float areaofcircle(float);
 float areaoftriangle(float, float);
 double sineofangle(double);
 double cosineofangle(double);
 double tangentofangle(double);
```

```
};
float myfirstclass::areaofcircle(float radius)
{
   answer = 3.14f * (pow(radius,2));
      return answer;
}
float myfirstclass::areaoftriangle(float b, h)
{
  answer = .5f * (b * h);
      return answer;
}
double myfirstclass::sineofangle(double angle)
{
  danswer = sin(angle);
      return danswer;
}
double myfirstclass::cosineofangle(double angle)
{
   danswer = cos(angle);
      return danswer;
}
double myfirstclass::tangentofangle(double angle)
{
  danswer = tan(angle);
      return danswer;
}
```

Step 2: Enter this code into your favorite text editor and save it as *10-2.cpp.*

```
#include "010-02.h"
#include <iostream>
using namespace std;
// function prototype
void menu();
// an instance of our class
myfirstclass myfirstobject;
int main ()
{
 menu();
    return 0;
  }
    void menu()
    {
```

```
double danswer, angle;
    float radius, base, height, area;
int imenu;
    cout << "Geometry Menu \n \n";
    cout << "1. Area of a Triangle \n";
cout << "2. Area of a Circle \n";
    cout << "3. Sine of an angle \n";
    cout << "4. Cosine of an angle \n";
    cout << "5. Tangent of an angle \n";
    cout << "6. Exit \n";
    cout << "Please enter the number of your selection
\n";
    cin >> imenu;
    switch(imenu)
    {
      case 1:
    cout << "Please enter the base of the triangle \n";
        cin >> base;
        cout << "Please enter the height of the triangle
\n";
        cin >> height;
        area =myfirstobject.areaoftriangle(base, height);
        cout << "The area of that triangle is " << area <<
"\n";
        cout << " \n";
        menu();
        break;
      case 2:
    cout << "Please enter the radius of the circle\n";
        cin >> radius;
    area = myfirstobject.areaofcircle(radius);
        cout << "The area of that circle is " << area <<
"\n";
        cout << " \n";
        menu();
        break;
      case 3:
    cout << "Please enter the angle in radians \n";
        cin >> angle;
        danswer =myfirstobject.sineofangle(angle);
        cout << "The sine of that angle is " << danswer <<
"\n";
        cout << " \n";
        menu();
```

```
        break;
      case 4:
    cout << "Please enter the angle in radians \n";
        cin >> angle;
        danswer = myfirstobject.cosineofangle(angle);
        cout << "The cosine of that angle is " << danswer <<
"\n";
        cout << " \n";
        menu();
        break;
      case 5:
    cout << "Please enter the angle in radians \n";
        cin >> angle;
        danswer =myfirstobject.tangentofangle(angle);
    cout << "The tangent of that angle is " << danswer
<< "\n";
        cout << "Press return to continue \n";
        cout << " \n";
        menu();
      case 6:
    return; // simply go back to main
        break;
      default:
     cout << "Please make a valid choice \n";
     menu(); // call this function again to display the
    // menu again
      }// end of switch
  }
```

Step 3: Compile the code and run it. You will see a series of images much like those depicted in Figures 10.2 and 10.3.

Clearly this particular program is a bit longer and more complex than most of the examples you have seen previously in this book. However, don't let that disturb you. The principles are the same as the simpler examples. We have taken some basic geometric and trigonometric functions and placed them in a class. That class is defined inside a header file. Instances of that class can be created inside any source code file where you may require the use of those functions. This previous example illustrates several points. To begin with, you see an example of the math functions found in math.h, as you had seen once before in a previous chapter. Also in this example we saw the use of a class, the creation and use of a header file, and finally the use of a menu function. This example is essentially the culmination of this book to this point and is of practical value,

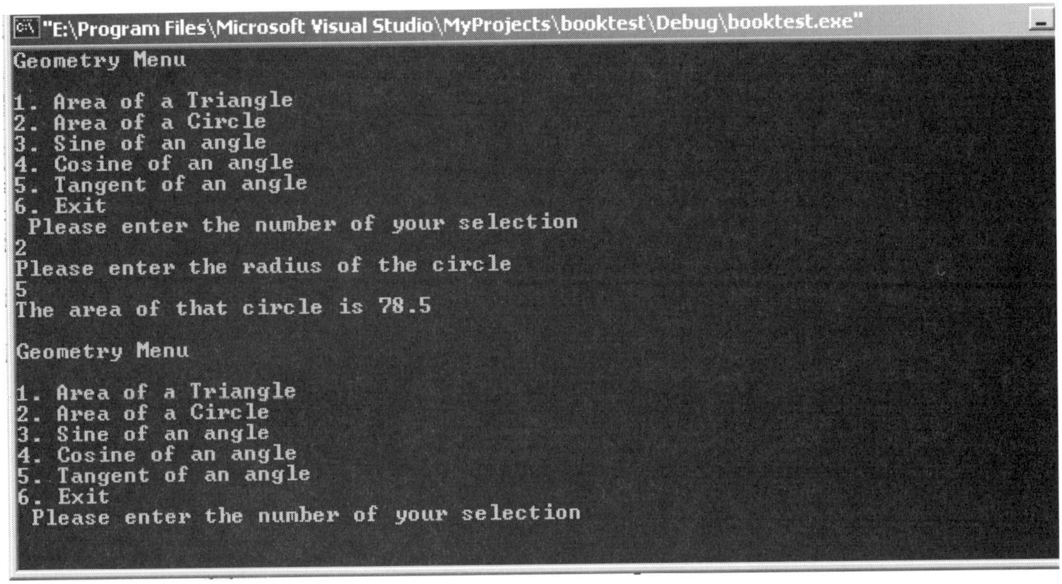

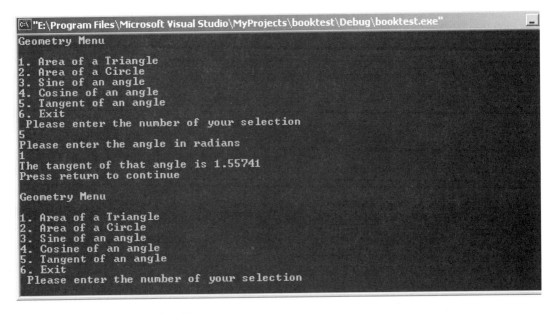

FIGURE 10.2   Geometry class.

FIGURE 10.3   Geometry class 2.

particularly to any readers currently engaged in the study of elementary geometry or trigonometry.

POINTERS TO CLASSES

As was thoroughly explained in Chapter 8, a pointer is just a reference to an address in memory. When you create an instance of a class, it must be stored in memory as well. This means that you can also create a pointer to an object created from a class.

HINT!

Remember that a class is just a template for creating an object; you have to make an instance of the class before you can use it.

A pointer to a class is done exactly the same way a pointer to a structure is. Remember that a class is really just a structure with functions in it. This means that when you access members of a pointer to a class you use the -> operator, just as you do with pointers to structures. Also remember that, as with all pointers, you must initialize the pointer before using it. Consider the following example.

Example 10.3

Step 1: Enter the following code into your favorite text editor.

```
#include <iostream>
using namespace std;
class myfirstclass
{
public:
 int number;
 void greeting();
};
void myfirstclass::greeting()
{
 for(int i = 0; i<= number;i++)
 {
```

```
        cout<< "Hello World \n";
    }
}
int main ()
{
    myfirstclass myfirstobject;
    myfirstclass *ptrclass;
    ptrclass= &myfirstobject;
    ptrclass->number = 5;
    ptrclass->greeting();
    return 0;
}
```

Step 2: Compile that code.

Step 3: Run the code. You should see something like what is shown in Figure 10.4.

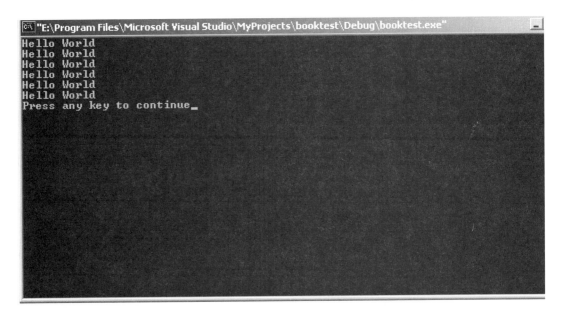

FIGURE 10.4   Pointers to classes.

All the classes you have seen in this chapter illustrate one of the four principle concepts of object orientation, and that is encapsulation. They all take the data to be stored and the functions that will manipulate that data, and put them together into a single class. This is the cornerstone of object-oriented programming. The principles of inheritance and polymorphism will be explored in more depth in the next two chapters.

CLASSES AND EXCEPTION HANDLING

Classes are quite useful and are the cornerstone for C++ programming. One way that many programmers utilize classes is by creating their own specialized classes to handle various exceptions. To do this, a programmer simply creates a class that has the appropriate methods to handle a particular error type. If you think back to Chapter 6, on error handling, you can throw various types of exceptions depending on the situation. This can be useful, particularly if you place these exception classes into a header file, then you can then use them in any program you wish. The following example shows the use of specialized exception classes.

Example 10.4

Step 1: Enter the following code into your favorite text editor and save it as *customexception.h*.

```
#include <fstream>
#include <iostream>
using namespace std;
class divisionbyzero
{
    fstream myfile;
public:

    void errormessage();
    void logerror();
};
class filenotfound
{
    fstream myfile;
public:
    void errormessage();
    void logerror();
};
void divisionbyzero::errormessage()
{
```

```
 cout << "Division by zero is not allowed \n";
}
void divisionbyzero::logerror()
{
myfile.open("error.txt",ios::app);
myfile<< "Division by zero error \n";
myfile.close();
}
void filenotfound::errormessage()
{
 cout << "File not found \n";
}
void filenotfound::logerror()
{
myfile.open("error.txt",ios::app);
myfile<< "File not found error \n";
myfile.close();
}
```

Step 2: Enter the following code into your favorite text editor and save it as *010-04.cpp*.

```
#include "test.h"
#include <iostream>
using namespace std;
// global variables
divisionbyzero d;
int main()
{
 int answer, dividend, divisor;
 try
{
  cout << "Please enter an integer \n";
  cin >> dividend;
  cout << "Please enter another integer \n";
  cin>> divisor;
  if(divisor ==0)
     throw d ;
      answer = dividend/divisor;
      cout << "The answer is " << answer;
}
catch(divisionbyzero de)
{
      de.errormessage();
```

```
        de.logerror();
  }
  return 0;
     }
```

Step 3: Compile the code.

Step 4: Run the program. You should see something much like what is shown in Figure 10.5.

```
Hello World
Hello World
Hello World
Hello World
Hello World
Hello World
Press any key to continue_
```

FIGURE 10.5 Creating your own exception classes.

HINT!

 You may have noticed that the functions for each of the specialized exception classes are repeated. This is a rather inelegant situation and in Chapter 11 you will see a way to correct this.

As you can see, creating such specialized exception classes allows you to handle different exceptions in a variety of ways. You simply include the header file that has the custom exception classes that you created, then you create a global instance of that class. Now you can use the methods of that class anywhere you wish. Also notice that the file object was private to both classes. This is because that object is only used inside the logerror method and there is no need for anyone to have external access to it. This gives you a great deal of flexibility in your exception handling.

CONSTRUCTORS AND DESTRUCTORS

Remember that a class is a rather complex data type. You will frequently find that there are some initial settings you might want to establish before a class is used. You might also need to retrieve data from some flat file or even a relational database. Most of this should be done before any methods of the class are called. You can do this in a constructor. A *constructor* is a function that has the exact same name as the class and is called when you instantiate the class. Constructors do not have any return type at all, not even void. Constructors can be very useful for setting initial values for certain member variables. Example 10.5 should help you to understand this.

Example 10.5

Step 1: Create a plain text file named input.txt that has exactly what you see here:
"Hello C++ Programmers."
Step 2: Please enter the following code into your favorite text editor.

```
#include <fstream>
#include <iostream>
using namespace std;
class myfirstclass
{
private:
     char msg[20];
     int loopcounter;
     fstream myfile;

public:
 void greeting();
 myfirstclass();
};
myfirstclass::myfirstclass()
{
     //open the file with the data
     myfile.open("input.txt",ios::in);
     myfile.getline(msg,20);
}
```

```
void myfirstclass::greeting()
{
     cout << msg << "\n";
}
int main ()
{
    myfirstclass myfirstobject;
    myfirstobject.greeting();
  return 0;
}
```

Step 3: Compile the code.
Step 4: Run the code. You should see something similar to Figure 10.6.

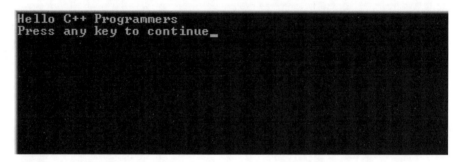

FIGURE 10.6   Constructors.

As you can see, the data printed-out to the screen is the same as what was retrieved from the text file. The constructor opened that file and put the data into the appropriate variables. Doing this in the constructor ensured that there was zero chance of trying to use one of the methods before the data was loaded into the variables.

Now we have a function that is called as soon as a class is instantiated. It can be used to initialize variables and retrieve data. But what about clean up? If we opened a flat file where do we close it? If we accessed a relational database, where do we close that connection? The answer is a *destructor*. A destructor is called when you destroy an object. A destructor is created much like a constructor, except the name is preceded by the negation symbol ~. The destructor will be executed as soon as the instance of the class goes out of scope—the end of the function in which the object was created. The following example illustrates this.

Example 10.6

Step 1: Please enter the following code into your favorite text editor.

```cpp
#include <fstream>
#include <iostream>
using namespace std;
class myfirstclass
{
private:
     char msg[20];
     int loopcounter;
     fstream myfile;

public:
 void greeting();
 myfirstclass();
 ~myfirstclass();
};
myfirstclass::myfirstclass()
{
     //open the file with the data
     myfile.open("input.txt",ios::in);
     myfile.getline(msg,20);
}
myfirstclass::~myfirstclass()
{
myfile.close();
}
void myfirstclass::greeting()
{
     cout << msg << "\n";
}
int main ()
{
     myfirstclass myfirstobject;
     myfirstobject.greeting();
     return 0;
}
```

Step 2: Compile the code.

The destructor closed the file for us, ensuring that we don't leave a file handle, or any other resource, still in memory. It is important that C++ programmers clean up after themselves.

If you will recall from the chapter on functions, it is possible to overload a function. An overloaded function occurs when you have multiple functions with the same name, but with different parameters. Overloaded functions can have either different numbers of parameters, or different types of parameters. You can overload constructors as well. The constructor that takes no arguments is the default constructor. Let's consider an example that will illustrate this for you.

Example 10.7

Step 1: Enter the following code into your favorite text editor.

```
#include <fstream>
#include <cstring>
#include <iostream>
using namespace std;
class myfirstclass
{
 private:
      char msg[20];
      int loopcounter;
      fstream myfile;

 public:
  void greeting();
  myfirstclass(char greeting[20]);
  myfirstclass();
  ~myfirstclass();
};
myfirstclass::myfirstclass()
{
     //open the file with the data
     myfile.open("input.txt",ios::in);
     myfile.getline(msg,20);
}
myfirstclass::myfirstclass(char greeting[20])
{
 memcpy(msg,greeting,sizeof(msg));
}
myfirstclass::~myfirstclass()
{
 myfile.close();
}
```

```
void myfirstclass::greeting()
{
     cout << msg << "\n";
}
int main ()
{
 myfirstclass myfirstobject("Howdy from Texas!    ");
     myfirstobject.greeting();
     return 0;
}
```

Step 2: Compile the code.

Step 3: Run the code. You will see something like what is depicted in Figure 10.7.

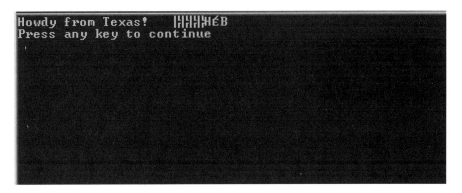

FIGURE 10.7   Multiple constructors.

As you can see in this example, you have two constructor functions. C++ determines which constructor to use by how you call it. If you create an instance of the object without passing it any parameters, then it will call the default no-argument constructor. If you pass it a character array, then it will call the constructor that takes a character array argument. If you pass it some other argument such as an int or a float, you will get an error.

Also notice the use of the memcpy function. This was first introduced in the chapter on functions. It copies contents from one array or structure (in our case, a character array) to another. The last parameter of that function is telling it how much to copy. Recall that you can enter a number and designate a certain number of bytes. However in this case, we used another function, `sizeof`, and told the function to copy a number of bytes equal to the size of the `msg` array.

ARRAYS AND POINTERS WITH CLASSES

It is important that you realize one important fact, once you have created a class, it is really a new data type, much like the existing data types (`int`, `float`, `bool`, etc.). Clearly the class is much more complex than an `int` or `float`, but it is still just a data type, and anything you can do with a standard data type you can do with a class. This means that you can create arrays of classes and pointers to classes. The following example shows the use of an array of classes. This example also uses a pointer to the array of classes, to input the data into the class.

HINT!

The array of classes in this example is rather small, that makes it easier for you to input data.

Example 10.8

Step 1: Write the following code into your favorite text editor.

```
#include <iostream>
using namespace std;
class student
{
public:
 char lastname[30];
 char firstname[30];
 float gpa;
 void displaydata();
};

void student::displaydata()
{
 cout << "*****Student info*****\n";
 cout << "lastname: " << lastname << endl;
```

```
  cout << "firstname:" << firstname << endl;
  cout << "GPA: " << gpa << endl;
}
int main()
{
student mystudents[5];
student *ptr = mystudents;
int i;
for (i = 0;i<5;i++)
{
  cout << "Please enter the students last name \n";
  cin>>ptr->lastname;
      cout << "Please enter the students first name \n";
  cin >> ptr->firstname;
      cout << "Please enter the students gpa \n";
  cin >> ptr->gpa;
      ptr++;
}// end of for loop

for(i=0;i<5;i++)
{
  mystudents[i].displaydata();
}
return 0;
}// end of main
```

Step 2: Compile and execute the code. You should see something much like what is depicted in Figure 10.8.

An array of classes is not much different from an array of integers. All the rules that normally apply to arrays still apply. The same is true of pointers. When you create a class, it will occupy a certain amount of memory. When you create a pointer to a class, it points to the first byte of memory occupied by that class.

One final note to remember is the keyword: this. In most object-oriented programming languages, this refers to the current instance of a class you are in. If you used code such as

```
this.funca();
```

you would be calling funca() inside of the current instance of the class. This keyword will be very important when you move on to *Visual C++*.

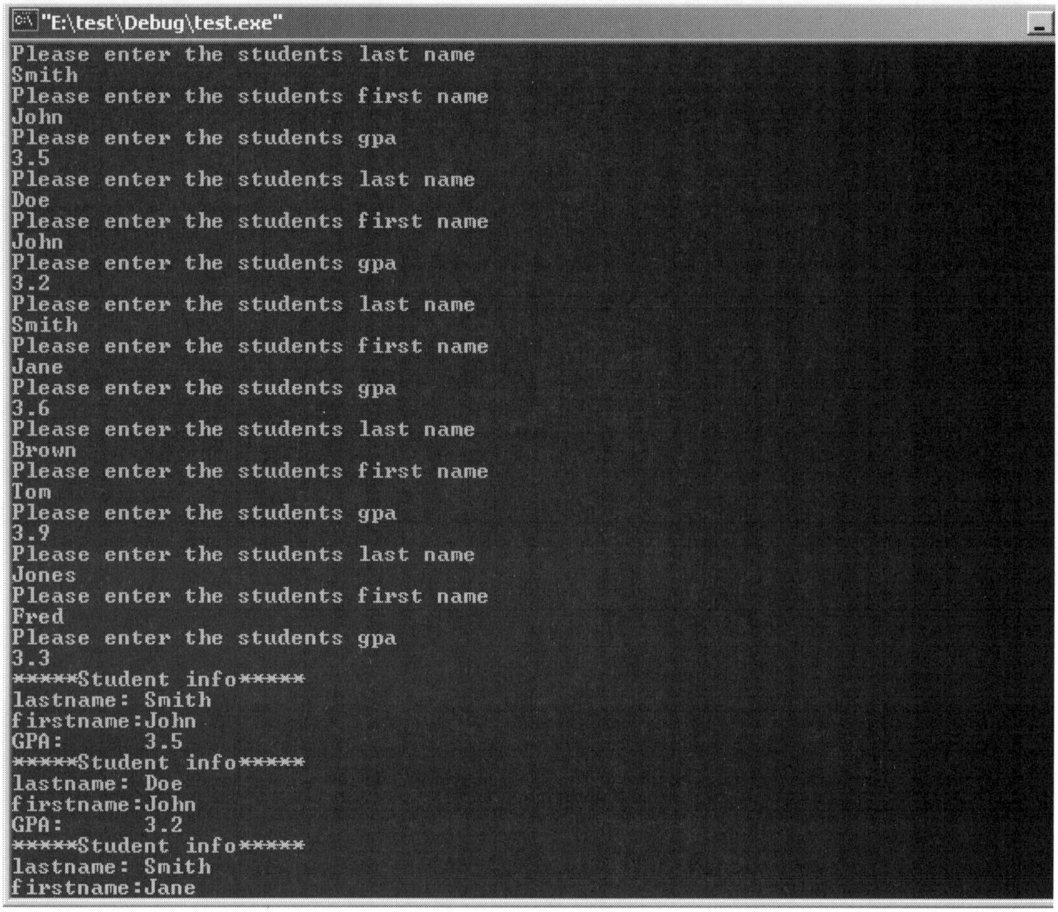

```
"E:\test\Debug\test.exe"
Please enter the students last name
Smith
Please enter the students first name
John
Please enter the students gpa
3.5
Please enter the students last name
Doe
Please enter the students first name
John
Please enter the students gpa
3.2
Please enter the students last name
Smith
Please enter the students first name
Jane
Please enter the students gpa
3.6
Please enter the students last name
Brown
Please enter the students first name
Tom
Please enter the students gpa
3.9
Please enter the students last name
Jones
Please enter the students first name
Fred
Please enter the students gpa
3.3
*****Student info*****
lastname: Smith
firstname:John
GPA:      3.5
*****Student info*****
lastname: Doe
firstname:John
GPA:      3.2
*****Student info*****
lastname: Smith
firstname:Jane
```

FIGURE 10.8   Pointers and arrays with classes.

SUMMARY

This chapter introduced you to the real reason for the development of C++. It is essentially object-oriented programming added to the framework of C programming. You have seen classes used to create objects, and you have seen those objects used within programs. This chapter has also introduced you to the basic concepts of object-oriented theory. The most important thing to remember is that, despite what may seem like archaic terminology, object-oriented programming is actually more natural than procedural programming. You should also remember that a class is a template for creating objects. When you create a class, you have essentially created your own complex data type, but one with its own methods. These concepts will be vital for the rest of this book.

REVIEW QUESTIONS

1. What are the four principles of object-oriented theory?
2. What is encapsulation?
3. What is the relationship between structures and classes?
4. What is the greatest advantage with object-oriented programming?
5. Why are classes frequently defined in header files?
6. What is inheritance?
7. Give a definition for a class.
8. What is the term for a function that is part of a class?
9. What is the term for a variable that is part of a class?

Inheritance

IN THIS CHAPTER
• • • • • • • • • • • • •

- Fundamentals of Inheritance
- Inheritance and Exception Handling
- Nested Classes
- Class Relationships
- Virtual Functions

Chapter 9 introduced you to object orientation and explored the concepts of abstraction and encapsulation in some depth. However, you have still not seen the most powerful and intriguing aspect of object-oriented programming, that of inheritance. *Inheritance* is a process whereby one class can inherit, or receive, the public and protected members of another class. Via inheritance you can create several similar classes without having to repeat the code that they have in common. You will see, in this chapter, that C++ brings you the full power in inheritance. This is important because some programming languages only give partial support for inheritance.

FUNDAMENTALS OF INHERITANCE
• •

Inheritance allows you to avoid replicating code unnecessarily. If you have some function or variable that is shared by more than one class, then you can declare that function or variable in one class, then have the others inherit that class. For example, if you needed to create a program that tracked animals in the zoo, you would have to create classes for all the animals that might be in the zoo. However, all animals have at least some things in common. Properties such as height and methods such as eat and procreate are common to all animals. So would it not make sense for you to create a class that has these methods and properties, then have the classes for the other animals inherit from this class? This is clearly demonstrated in Figure 11.1.

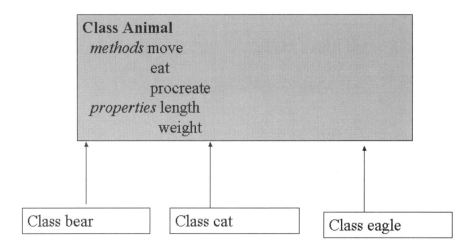

FIGURE 11.1   Inheritance hierarchy.

This concept is really not that complicated. You simply create a class that has the properties and methods that will be shared by multiple classes, then the other classes inherit from this class. As with all things in programming, there is a specific set of terminology for inheritance. The class that is inherited from can be referred to by a number of terms, the most common being *parent class* or *base class*. The class that does the inheriting is referred to as the *child class* or the *derived class*. The way you inherit from a class in C++ is to list the name of your class and then use the scope resolution operator (a single colon :) followed by the name of the class you wish to inherit from. The following example will show you how this works.

Example 11.1

Step 1: Enter the following code into your favorite text editor and save it as *11-1.h*.

```
    class bankaccount
    {
protected:
    float balance;
public:
    float withdraw(float);
    float deposit(float);
    void displaybalance();
    bankaccount();
};
bankaccount::bankaccount()
{
 balance = 1100;
}
float bankaccount::withdraw(float amount)
{
balance -= amount;
    return balance;
}
float bankaccount::deposit(float amount)
{
 balance += amount;
    return balance;
}
void bankaccount::displaybalance()
{
 cout << "Your balance is " << balance << endl;
}
class checking:public bankaccount
{
};
class savings:public bankaccount
{
};
```

Step 2: Enter the following code into your favorite text editor and save it as *11-1.cpp*.

```cpp
#include "test.h"
#include <iostream>
using namespace std;
// function prototypes
void menu();
// declare an instance of the checking class
checking mychecking;
int main()
{
 menu();
 return 0;
}
void menu()
{
 // declare other variables
 int menuchoice;
 float amount, balance;
 // display choices
 cout << "****** Acme Bank Checking Account*******"<<
endl;
 cout << endl;
 cout << "1. Check Balance "<< endl;
 cout << "2. Make Deposit "<< endl;
 cout << "3. Make withdrawal " << endl;
 cout << "4. Exit " << endl;
 cout << "Please enter your choice " << endl;
 // input selection
 cin >> menuchoice;
 // do a switch based on selection
 switch(menuchoice)
 {
    case 1:
mychecking.displaybalance();
      cout << endl;
menu();
      break;
    case 2:
      cout << "Please enter the amount to deposit \n";
      cin>> amount;
balance = mychecking.deposit(amount);
      cout << "New balance is " << balance << endl;
      cout << endl;
menu();
      break;
```

```
    case 3:
      cout << "Please enter the amount to withdraw \n";
      cin >> amount;
balance = mychecking.withdraw(amount);
      cout << "New balance is " << balance << endl;
      cout << endl;
menu();
      break;
    case 4:
return;
      break;
    default:
      cout << "Your choice was invalid \n";
      menu(); // re display the menu
  }// end of switch
}
```

Step 3: Compile this code. When you run the code, you should see
something like what is depicted in Figure 11.2.

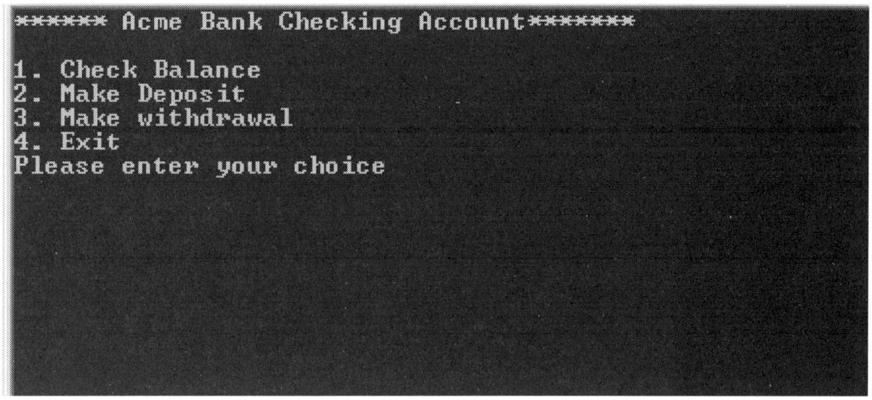

FIGURE 11.2   Inheritance.

You can now see how you can inherit methods and properties from the
parent class. This example also has practical value because it provides a
good base for creating an actual banking program. What is of particular
importance in this example is the fact that you can clearly see how the
public methods that were inherited by the child class are now available to
the child class, as if they were part of that class. This is the entire purpose

for using inheritance. The definition of inheritance given earlier mentioned that it was when one class received a copy of another class's public and protected methods. You may be wondering what this means. Public and protected are both access modifiers. They determine how items outside of a class can access a method or property in a class. Table 11.1 shows the three basic access modifiers.

TABLE 11.1   Access Modifiers

Access Modifier	Purpose
Private	Items that are designated private cannot be accessed from outside the class they are in. They also cannot be inherited.
Public	Items that are public can be accessed from outside the class, and they can be inherited.
Protected	Items that are protected cannot be accessed from outside the class, but they can be inherited.

When you instantiate a derived class, its base class constructor will be called automatically. This is actually a good thing for you because constructors hold code that must be executed in order for the class to function properly. The following example illustrates this.

Example 11.2

Step 1: Enter the following code into a text editor and save it as *11-02.h.*

```
class baseclass
{
public:
      baseclass();
};
baseclass::baseclass()
{
 cout << "Base class constructor is executing \n";
}
class derivedclass:public baseclass
{
public:
     derivedclass();
```

```
};
derivedclass::derivedclass()
{
 cout << "Derived class constructor is executing \n";
}
```

Step 2: Enter this code into a text editor and save it as *11-02.cpp.*

```
#include "test.h"
#include <iostream>
using namespace std;
int main()
{
 derivedclass myclass;
 return 0;
}
```

Step 3: Compile and execute the code. You should see an image much like what is shown in Figure 11.3.

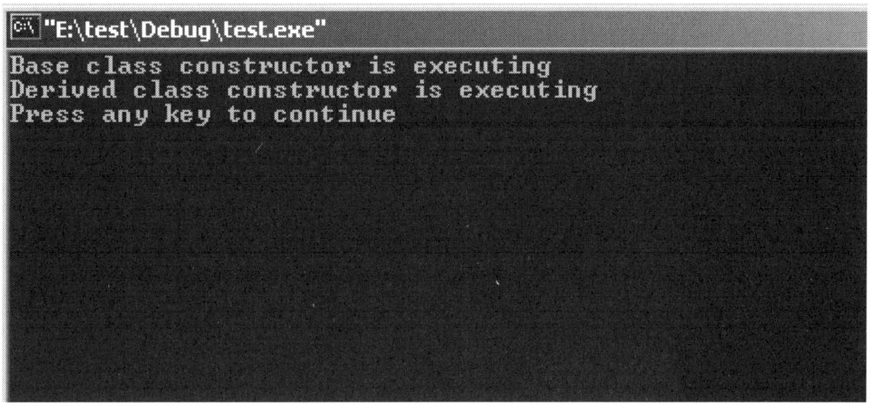

FIGURE 11.3   Calling the base classes, constructor.

This program should give you a simple understanding of the order in which constructors are called. It should also show you that the constructor is executed as soon as you instantiate the class.

An issue arises when your base class has multiple constructors and you wish your derived class to call one of the other constructors (i.e., not the default no arguments constructor). The way to do this is to modify the definition of your child class's constructor so that it will in turn call the appropriate base class constructor. You do this by following the derived class's constructor definition with a colon, and a call to the parent class constructor you wish to call. The following is an example of this.

```
checking::checking(float num):bankaccount(num)
{
}
```

Now if you pass a float number to the constructor of the checking account class, it will in turn call the constructor of the parent class that takes a float, and pass that along. The following example is a modification of the previous program that uses overloaded base class constructors.

Example 11.3

Step 1: Enter the following code into your favorite text editor and save it as *11-3.h*.

```
class bankaccount
{
 protected:
     float balance;
public:
     float withdraw(float);
     float deposit(float);
     void displaybalance();
     bankaccount(float);
     bankaccount();
};
bankaccount::bankaccount(float num)
{
   balance = num;
}
bankaccount::bankaccount()
{
 balance = 1100;
}
float bankaccount::withdraw(float amount)
{
 balance -= amount;
     return balance;
```

```
}
float bankaccount::deposit(float amount)
{
 balance += amount;
    return balance;
}
void bankaccount::displaybalance()
{
 cout << "Your balance is " << balance << endl;
}
class checking:public bankaccount
    {
public:
    checking(float);
    };
    checking::checking(float num):bankaccount(num)
    {

    }
    class savings:public bankaccount
    {
    };
```

Step 2: Enter the following code into your favorite text editor and save it as *11-3.cpp*.

```
#include "test.h"
#include <iostream>
using namespace std;
    // function prototypes
    void menu();
    // declare an instance of the checking class
    checking mychecking(1200);
    int main()
    {
menu();
return 0;
    }
    void menu()
    {
// declare other variables
int menuchoice;
float amount, balance;
// display choices
```

```
cout << "****** Acme Bank Checking Account*******"<<
  endl;
cout << endl;
cout << "1. Check Balance "<< endl;
cout << "2. Make Deposit "<< endl;
cout << "3. Make withdrawal " << endl;
cout << "4. Exit " << endl;
cout << "Please enter your choice " << endl;
// input selection
cin >> menuchoice;
// do a switch based on selection
switch(menuchoice)
{
case 1:
  mychecking.displaybalance();
      cout << endl;
 menu();
      break;
case 2:
      cout << "Please enter the amount to deposit \n";
      cin>> amount;
   balance = mychecking.deposit(amount);
      cout << "New balance is " << balance << endl;
      cout << endl;
   menu();
      break;
 case 3:
      cout << "Please enter the amount to withdraw \n";
      cin >> amount;
   balance = mychecking.withdraw(amount);
      cout << "New balance is " << balance << endl;
      cout << endl;
   menu();
      break;
case 4:
  return;
      break;
default:
      cout << "Your choice was invalid \n";
      menu(); // re display the menu
 }// end of switch
}
```

Step 3: Compile your code.

Step 4: Execute the code.

In short, what you are doing is telling the base class which constructor you wish to use. This can be done regardless of how many constructors the base class might have. All you need to do is specify which base class constructor to use when you define the constructors for the derived class. You may also encounter situations where the base class has multiple constructors. You can have any of the child class constructors call any of the base class constructors; however, the logical course usually involves matching the parameters of the child classes constructor with the base class constructor it will call. In other words, if your base class has a constructor that takes a float, then you will probably call that from a derived class constructor that takes a float and pass the float from the child class constructor to the parent class constructor.

INHERITANCE AND EXCEPTION HANDLING

In Chapter 9 you saw how to use custom exception classes to enhance your error handling. Exception classes are not too different from any other classes. You will probably have some methods that are common for all exception classes you create. For this reason, it makes sense to create a base class that contains the common elements that you wish all exception classes to have. You can then create specific classes for specific exception types and use them as needed.

HINT!

Some other programming languages such as *Java* include a host of such exception classes for you. With C++, you will need to create your own exception classes.

The following example illustrates the use of a base exception class and several derived classes. It is strongly suggested that you consider including the header file with these classes in all your programs so that you might use these exception classes in all the code you write.

Example 11.4

Step 1: Please enter the following code into your favorite text editor and save it in a file name *error.h*.

```cpp
#include <fstream>
#include <memory>
#include <iostream>
using namespace std;
class error
{
protected:
 fstream errfile;
public:
 void printmsg(char *msg);
 void logmsg(char *msg);
};
void error::printmsg(char *msg)
{
 cout << msg;
}
void error::logmsg(char *msg)
{
 errfile.open("error.log",ios::app);
 errfile << msg;
}
class fileerror:public error
    {
    };
    class divisionbyzero:public error
    {
    };
```

Step 2: Enter the following code into your favorite text editor and save it in a file named *11-4.cpp*.

```cpp
#include "error.h"
#include <iostream>
using namespace std;
fileerror fe;
divisionbyzero de;
int main()
{
    fstream f;
```

```
          float dividend, divisor,quotient;
      try
      {
       cout << "Enter a number " << endl;
          cin >> dividend;
          cout << "Enter another number " << endl;
          cin >> divisor;
          // if the divisor is zero we should
          // get the divisionbyzero
          if(divisor==0)
          throw de;
          quotient = dividend/divisor;
          cout << quotient;
          // if there is no file then we should get
          // the fileerror
          f.open("somefile.txt",ios::in);
          if(!f)
          throw fe;
      }
      catch(divisionbyzero d)
      {
      d.printmsg("Division by zero \n");
      }
      catch(fileerror f)
      {
      f.printmsg("File not found error \n");
      }
      catch(...)
      {
      cout<< "Unknown error! \n";
      }
          return 0;
        }
```

Step 3: Compile your code.

Step 4: Run the program. You should see something like what is depicted in Figure 11.4.

This example, hopefully, accomplishes several goals. To begin with it illustrates how you might create custom error classes and use inheritance to avoid unnecessarily repeating code. But it is also one more opportunity for you to practice inheritance techniques in C++. You can easily create any type of custom exception class you wish. You can also create the proper types of error functions to do anything you might wish to do

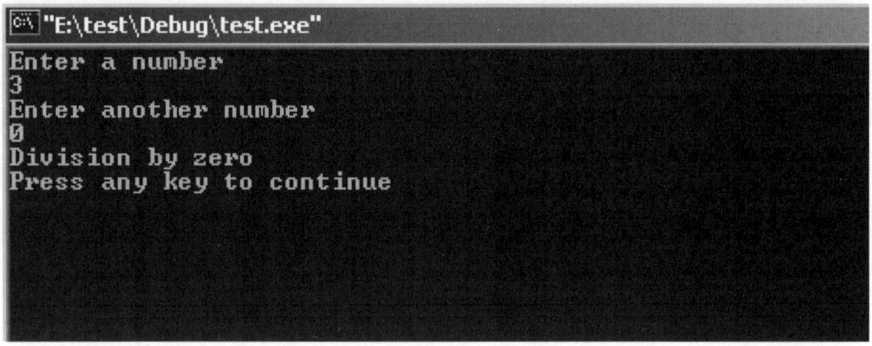

FIGURE 11.4   Custom exception classes.

to either log the error, or, perhaps, fix the error. You may be wondering how you go about fixing an error. The following example illustrates how you might fix the division by 0 error. The following code snippet adds a function to the division by 0 class, and this function takes the address of the dividend, and changes it to a non-0 number.

```
void fixzeroerror(float &dividend)
{
 dividend += 1;
}
```

Now, with this function added to the division by 0 class, all you need to do when a division by 0 error occurs is call this function. This function will change the dividend to a non-0 number.

NESTED CLASSES

One area we have not yet explored is the concept of nested classes. You can define a class within another class. You can also have a class that, within its functions, instantiates another class. In fact, you will frequently see the latter example of a nested class used in various programs. The first example, that of constructing a class totally within another class, is not as commonly used. That class, defined within another class, can only be used inside of that class. It cannot be executed from outside the class, even if you declare it as public. Let's take a look at a class declared within another class.

Example 11.5

Step 1: Write the following code in your favorite text editor and save it as *11-05.h.*

```
class baseclass
{
public:
 void basefunca();
     class nested
     {
       public:
     void nestedfunca();
     };
};
void baseclass::basefunca()
{
 cout << "This is a function in the base class\n";
 nested myclass;
 myclass.nestedfunca();
}
void baseclass::nested::nestedfunca()
{
 cout << "This is a function in the nested class \n";
}
```

Step 2: Write this code into a text editor and save it as *11-05.cpp.*

```
#include "11-05.h"
#include <iostream>
using namespace std;

int main()
{
 baseclass myclass;
 myclass.basefunca();

 return 0;
}
```

Step 3: Compile the code and execute the program. You should see something similar to what is depicted in Figure 11.5.

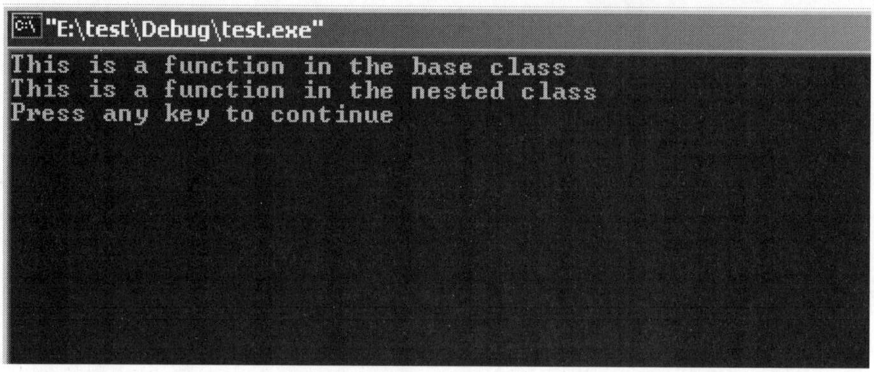

FIGURE 11.5 Simple inheritance.

You can test the claim that nested classes cannot be accessed from outside the class. Simply try any method you wish of creating an instance of that class from the main function. You will find that no method you try is effective. The purpose of creating a class within another class is so that the container class can use the methods and properties of the nested class.

As was mentioned earlier in this chapter, it is also possible to have a normal class but to create an instance of it within another class. This is similar to nested classes, with one key difference: with a nested class, it can only be used within the class it is defined. A normal class that is simply instantiated inside another class can still be used in the normal way. The following example illustrates this.

Example 11.6

Step 1: Type the following code into your favorite text editor and save it as *11-06.h.*

```
class classA
{
public:
 void funca();
};
void classA::funca()
{
 cout << "This is a function in classA \n";
}
```

```
class classB
{
public:
  void funcb();
};
void classB::funcb()
{
 classA myclass;
 myclass.funca();
}
```

Step 2: Type this code into a text editor and save it as *11-06.cpp*.

```
#include "11-06.h"
#include <iostream>
using namespace std;
int main()
{
classB myclass;
classA anotherclass;
myclass.funcb();
anotherclass.funca();

return 0;
}
```

Step 3: Compile your program and execute it. You should see an image much like what is shown in Figure 11.6.

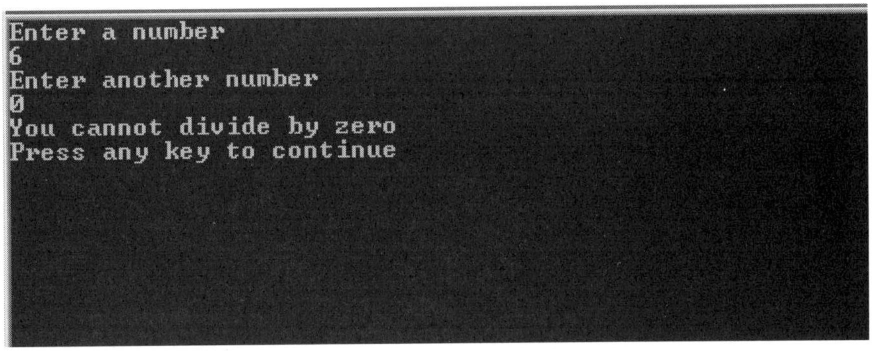

FIGURE 11.6   Instantiating a class within another class.

You can see in this example that the same function was called from within the containing class, and then again from simply making a direct instance of the contained class. This is the essential difference between instantiating a normal class within another class, and a truly nested class.

CLASS RELATIONSHIPS

When one class inherits from another, there is some inherent relationship between the classes. Defining what that relationship is, is just a matter of carefully considering how the classes are used. If a class inherits from another class and is simply a special case of that base class, then the derived class has an *is-a* relationship with the base class. Consider the case where you have a class called animal and another class called dog, which inherits from animal. You then have an *is-a* relationship. The class dog, is an animal.

When one class contains another, as in Examples 11.5 and 11.6, you have a *has-a* relationship. The practical example would be that if you had a class called car, that uses a class called engine, then the relationship would be a *has-a* relationship, the car has an engine.

VIRTUAL FUNCTIONS

You are probably wondering what a virtual function is. A virtual function is a function that must be overridden. You declare a virtual function in a base class when you want all the derived classes to have that function, but you want to force them to override it with their own specific implementation. Put another way, if you declare a virtual function in a base class, all the derived classes will inherit it, but they cannot use it as is. They must override it with their own implementation.

Declaring a function as virtual is very easy. Simply precede the function name with the keyword *virtual*. Perhaps an example would be helpful. For this example, we will work with an error class example with inheritance, and simply add a virtual function to it.

Example 11.7

Step 1: Open your favorite text editor and write in the following code. Save the file as *11-07.h*.

```
const int DIVISIONBYZERO = 1;
const int FILENOTFOUND =2;
//***********define the base class********************
class customerror
{
protected:
    int type;
public:
    int errortype();
    virtual void errormsg();
};// end of class definition
int customerror::errortype()
{
 return type;
}// end of error type
void customerror::errormsg()
{
 cout << "you will never see this because it will be
overridden!\n";

}// end of errormsg
//*******end of base class********************
class divisionbyzero: public customerror
{
public:
    divisionbyzero(int);
    void errormsg();
};//end of divisionbyzero class definition
divisionbyzero::divisionbyzero(int errortype)
{
    type = errortype;
}
void divisionbyzero::errormsg()
{
 cout << "You cannot have divide by zero \n";
}
class filenotfound: public customerror
{
public:
    filenotfound(int);
    void errmsg();
};
filenotfound::filenotfound(int errortype)
{
```

```
 type = errortype;
}
void filenotfound::errmsg()
{
 cout << "Cannot find that file!\n";
}
```

Step 2: Type this code into your favorite text editor and save it as *11-07.cpp*.

```
#include "customerror.h"
#include <fstream>
#include <iostream>
using namespace std;
int main()
{
 divisionbyzero de(DIVISIONBYZERO);
 filenotfound fe(FILENOTFOUND);
 float dividend, divisor, quotient;
 try
 {
  cout << "enter a number \n";
  cin >> dividend;
  cout << "enter another number \n";
  cin >> divisor;
      if (divisor==0)
      {
throw de;
      }
quotient = dividend/divisor;
cout << "The answer is "<< quotient << endl;
fstream myfile;
myfile.open("c:\\something.txt",ios::in);

if(!myfile)
{
 throw fe;
}
}// end of try
catch(divisionbyzero de)
{
de.errormsg();
```

```
}
catch(filenotfound fe)
{
fe.errormsg();
}// end of filenotfound catch
catch(...)
{
 cout << "Something bad happened \n";
}//end of final catch
return 0;
    }// end of main
```

Step 3: Compile and execute your code. You should see something like what is shown in Figure 11.7.

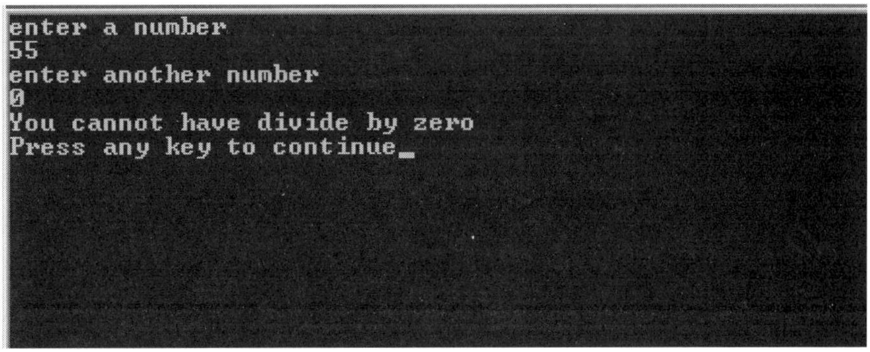

FIGURE 11.7   Virtual functions.

Now that you have seen how a virtual function works, you might be wondering why one would wish to use them. Could you not simply write individual functions in each of the derived classes and skip the virtual function in the base class? The answer is "yes, you could." However, the virtual function accomplishes a few goals. First, it forces anyone making a derived class from this base class to create an implementation of that function. This guarantees that such a programmer using classes derived from your base class cannot forget to implement this function. Furthermore, it ensures that all such functions will have the same basic declaration line.

SUMMARY

This chapter introduced you to inheritance. Inheritance is a powerful feature that object-oriented programming offers to you. Learning to use it is critical. Many modern software products have an object model hierarchy that you can access. These hierarchies are dependent on one object inheriting from another. In this chapter you also were introduced to the various relationships found in inherited classes. These relationships and the associated terminology are of great important in object-oriented programming.

REVIEW QUESTIONS

1. What is inheritance?
2. How do you call an overloaded constructor in the base class?
3. What is another word for base class?
4. How would you declare class a so that it inherits from class b?
5. What is another term for derived class?
6. What is a protected variable?
7. What is another word for a function that is part of a class?
8. Are private members inherited?
9. When you have a derived class, which constructor is executed first . . . the derived class constructor or the base class constructor?
10. When one class contains another class, what type of relationship is there?

12

Advanced
Object-Oriented
Concepts

IN THIS CHAPTER
• • • • • • • • • • • • • •

- Polymorphism
- Multiple Inheritance
- Indirect Inheritance
- Pointers to Classes
- Abstract Classes and Pure Virtual Functions

Chapter 11 introduced you to inheritance, the process whereby one class gets a copy of the public and protected methods of another class. In the previous two chapters, you have seen three of the four principle concepts of object-oriented theory. You have seen abstraction, encapsulation, and inheritance. This chapter will show you how to use the fourth principle, that of polymorphism. The word *polymorphism* literally means "many forms." It simply means that once you have inherited a function, you can override the function you inherited and change it.

POLYMORPHISM

When a class inherits a base class, it receives all the base class's public and protected methods and properties. However, it does not have to take everything it gets as is. The derived class can override any function to suit the needs of the derived class. In other words, you are not stuck with the functions you inherit. You can override them. Overwriting a function is not the same as overloading a function. When you overload a function it must have different parameters (either a different number or different types). When you override a function it should have the same parameters, but the actual code in the function will be different. The function declaration (the name, return type, and parameters) is called the *interface*; the actual code in the function is called the *implementation*. Thus, your interface will be the same when you override an inherited function, but its implementation will be the same. The following example will clarify this.

Example 12.1

Step 1: Enter the following code into you favorite text editor. Save it as *12-01.h.*

```
class baseclass
{
public:
     void testfunction();
};
void baseclass::testfunction()
{
 cout << "Hey this is the base class!\n";
}
class childclass:public baseclass
{
public:
     void testfunction();
};
void childclass::testfunction()
{
 cout << "Hey this is in the child class!!\n";
}
```

Step 2: Enter this code into your favorite text editor and save it as *12-01.cpp.*

```
#include "test.h"
#include <iostream>
using namespace std;
int main()
{
 childclass myclass;
 myclass.testfunction();
return 0;
}
```

Step 3: Compile your code.

Step 4: Run your application, you should see something similar to what is shown in Figure 12.1.

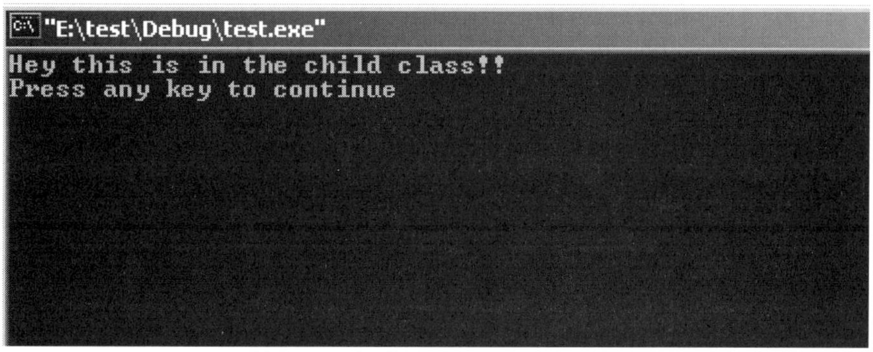

FIGURE 12.1   Overwriting an inherited function.

This example illustrates the basics of polymorphism. You can change the form of any function you inherit. As you can see in Example 12.1, the function that is called is the one in the derived class, not the one in the base class.

Although this does show you the basics of polymorphism, it does not show you the power of polymorphism. This comes into play when you have many derived classes, and each needs to implement a particular method in a slightly different way. Then, each class can overload the method inherited and have the method do what they wish. Consider this example.

Example 12.2

Step 1: Enter the following code into your favorite text editor and save it as *12-02.h.*

```cpp
class baseclass
{
public:
      void computetax(float);
};
void baseclass::computetax(float price)
{
 cout << "Hey this is the base class!\n";
 cout << price * .075f;
}
class childclass1:public baseclass
{
public:
      void computetax(float);
};
void childclass1::computetax(float price)
{
 cout << "Hey this is the child class 1\n";
 cout << price * .010f << endl;
}
class childclass2:public baseclass
{
public:
 void computetax(float);
};
void childclass2::computetax(float price)
{
 cout << "Hey this is the child class 2\n";
 cout << price * .005f << endl;
}
```

Step 2: Enter this code into your favorite text editor and save it as *12-02.cpp.*

```cpp
#include "test.h"
#include <iostream>
using namespace std;
```

```
int main()
{
 childclass1 myclass;
 childclass2 anotherclass;
 myclass.computetax(100);
 anotherclass.computetax(100);
 return 0;
}
```

Step 3: Compile the code.

Step 4: Run the code, you should see something much like what is shown in Figure 12.2.

```
"E:\test\Debug\test.exe"
Hey this is the child class 1
1
Hey this is the child class 2
0.5
Press any key to continue
```

FIGURE 12.2   Polymorphism with multiple derived classes.

As you can see, each of the child classes has a separate implementation for the function computetax. This is how polymorphism is generally used. You have different classes with a function of the same name, and even the same parameters, but with different implementations.

HINT!

You only need to redefine a function in the derived class if you intend to override it. If you will use it as it is in the base class then you do not need to take any action at all.

MULTIPLE INHERITANCE

C++ offers complete support for inheritance, unlike some other object-oriented programming languages. Specifically this means that C++ supports multiple inheritance. This means that a derived class can inherit from more than one base class. This is not nearly as complicated as it sounds. To inherit multiple base classes, you simply separate them with commas as you see in the following example.

```
class derivedclass:public baseclass1,public baseclass2,
    public baseclass3
{
}
```

The following example will show you that multiple inheritance is not much different from single inheritance.

Example 12.3

Step 1: Enter the following code into your favorite text editor and save it as *12-03.h*.

```
class baseclass1
{
public:
      float computetax(float);
};
float baseclass1::computetax(float amount)
{
return amount * .075f;
}
class baseclass2
{
public:
      float computededuction(float);
};
float baseclass2::computededuction(float amount)
{
return amount - (amount *.10f);
}
class derivedclass:public baseclass1,public baseclass2
{
};
```

Step 2: Enter this code in your favorite text editor and save it as *12-03.cpp*.

```
#include "12-03.h"
#include <iostream>
using namespace std;
int main()
{
 derivedclass myclass;
 float answer, amount;
 cout <<"Enter the amount to tax \n";
 cin>> amount;
 answer = myclass.computetax(amount);
 cout << "The tax is " << answer << endl;
 cout << "Enter the amount to reduce \n";
 cin >> amount;
answer = myclass.computededuction(amount);
cout << "The amount after deduction is "<< answer <<
endl;
return 0;
}
```

Step 3: Compile the code.

Step 4: Run the code. You should see something like what is shown in Figure 12.3.

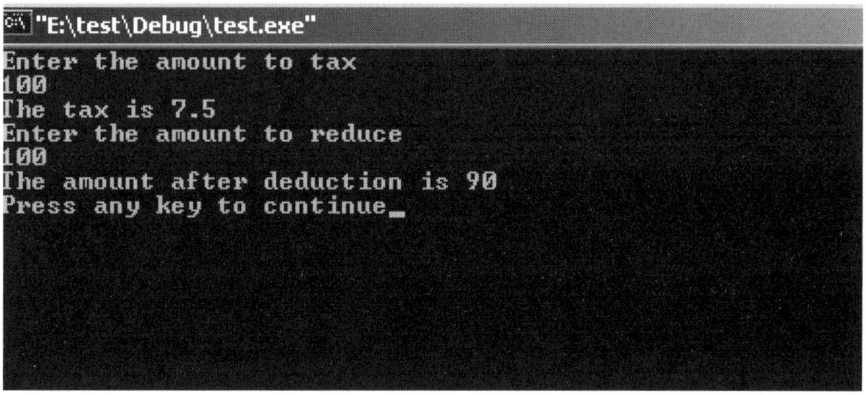

FIGURE 12.3   Multiple inheritance.

As you can see in this example, you can use multiple inheritance to get functions you need from more than one base class. The entire cornerstone of object-oriented programming is code reusability. This can only be realized through the use of inheritance, both standard and multiple.

Multiple inheritance can be a powerful feature but it does pose some unique problems. Assume that you are inheriting from base class A and base class B, and both have a method called funca(). When your derived class calls funcA(), which function is it calling? The answer to this unique problem is to simply override any functions that both base classes have in common. Basically you use the techniques covered in Section 1 to override the function that is duplicated. The following example is a bit more complex and illustrates handling multiple inherited functions.

Example 12.4

Step1: Enter the following code into your favorite text editor and save it as *12_04.h*.

```
class bankaccount
{
protected:
     float balance;
     float penalty;// for overdraft
     float interestrate;
     float fee; // for various transactions
public:
     float withdraw(float);
     float deposit(float);
     bankaccount();
};
bankaccount::bankaccount()
{
     balance =1000.00f;
     penalty =25.00f;// for overdraft
     interestrate=.02f;
     fee =10.00f; // for various transactions
}
float bankaccount::deposit(float amount)
```

```
{
balance -= amount;
return balance;
}
float bankaccount::withdraw(float amount)
{
balance +=amount;
return amount;
}
class transferfunds
{
protected:
      float fee; // transfer fee
public:
      bool transfermoney(bankaccount *source,bankaccount
 *dest,float);
   transferfunds();
};
transferfunds::transferfunds()
{
   fee = 20.00f;
}
bool transferfunds::transfermoney(bankaccount
 *source,bankaccount *dest,float amount)
   {
 source->withdraw(amount);
 dest->deposit(amount);
 return true;
   }
class checking:public bankaccount,public transferfunds
{
 public:
      void deductmonthlyfee();
};
void checking::deductmonthlyfee()
{
balance -= fee;
}
class savings:public bankaccount,public transferfunds
{
};
```

HINT!

You see that both the bankaccount class and the transfer-funds class have a variable called fee. When other classes inherit from both bankaccount and transferfunds, they must override that variable. This is the problem with multiple inheritance. If you compile this code, as you see it you will get an error because the compiler cannot tell which fee to use in the deductfee function. Thus, we will override fee in the checking class and add a constructor to the checking class. It is essential to prototype it as well. The following is the constructor.

```
checking::checking()
{
fee = 5.00f;
}
```

You probably noticed the function transfermoney, which takes some parameters you have not seen before. These are pointers to classes, which will be explained in detail later in this chapter.

Step 2: Enter the following code into your favorite text editor and save it as *12_04.cpp.*

HINT!

Whereas the header file demonstrates the multiple inheritance that we wanted to examine, the code we are about to look at provides a more practical, functional example with a scaled-down bank account application.

```
#include "12-03.h"
#include <iostream>
using namespace std;
void menu();
int main()
{
    void menu();
    return 0;
}// end main
```

```
void menu()
{
 int imenu;
 char acct;
 float amount, balance;
 savings mysavings;
 checking mychecking;
 cout << "1. Deposit funds \n";
 cout << "2. Withdraw funds \n";
 cout << "3. Exit \n";
 cout << "Please enter your choice \n";
 cin >> imenu;
 switch(imenu)
 {
  case 1:
    cout <<"From checking (c) or savings (s) \n";
       cin >> acct;
    cout << "How much ? \n";
      cin >> amount;
       if(acct == 'c')
       balance = mychecking.withdraw(amount);
       else
       balance = mysavings.withdraw(amount);
        cout << "The new balance is " << balance << endl;
        break;
 case 2:
   cout <<"From checking (c) or savings (s) \n";
       cin >> acct;
    cout << "How much ? \n";
       cin >> amount;
       if(acct == 'c')
       balance = mychecking.deposit(amount);
        else
       balance = mysavings.deposit(amount);
        cout << "The new balance is " << balance << endl;
        break;
  case 3:
       return;
 default:
       cout << "Invalid choice \n";
}//end switch
 menu();
   }// end menu
```

If you take the time to carefully examine this code you will see that much of the code consists of techniques we have previously examined. The real item here we wanted to examine involved overriding multiple inherited variables; this was done in the header file. However, the program provided gives you a basic scaled-down bank account program. With a little thought and time, you could easily grow this into a full-fledged banking account program.

INDIRECT INHERITANCE

In addition to multiple inheritance, you will also encounter situations where you will inherit from a class that is, in turn, inheriting from another class. In fact, later in this book, when we get to *Visual C++*, you will see a lot of this. Essentially if your class is derived from a class, you inherit all its protected and public members, even if those members were inherited from yet another class. You can have inheritance going back any number of classes and you will have inherited it all.

Example 12.5

Step 1: Enter the following code into your favorite text editor.

```
#include <iostream>
using namespace std;
class baseclass1
{
 public:
    void basefunc1();
};
void baseclass1::basefunc1()
{
 cout << "This is in the first base class \n";
}
class baseclass2: public baseclass1
{
 public:
    void basefunc2();
};
void baseclass2::basefunc2()
{
```

```
      cout << "This is in the second base class \n";
}
class baseclass3: public baseclass2
{
public:
  void basefunc3();
};
void baseclass3::basefunc3()
{
  cout << "This is in the third base class\n";
}
class derivedclass : public baseclass3
{
};
int main ()
{
 derivedclass myclass;
  myclass.basefunc1();
  myclass.basefunc1();
  myclass.basefunc3();

  return 0;
}
```

Step 2: Compile and execute your program.
You should see something much like what is depicted in Figure 12.4.

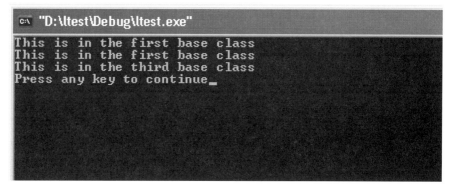

FIGURE 12.4   Indirect inheritance.

You can see that the derived class you actually used has access to all the public functions it inherited from each class. This example shows several layers of indirect inheritance.

POINTERS TO CLASSES

As you have previously seen, you can make a pointer to any data type. And as we have already discussed, a class is simply a data type, it's just a lot more complicated than most. This means you can use pointers to classes. You have already seen an example shown with the transfer-money function. Recall that the purpose of a class is to take the data and functions that work on the data and encapsulate them together. If you require a function to have access to both data and functions, it is only logical to pass it a pointer to a class.

The pointer to a class operates just like the pointer to a structure. You use the -> operator rather than the . operator to access the various members of the class pointer. The following brief code segment illustrates this.

```
void funca(myclass *myobj)
{
myobj->element1;
myobj->element2;
}
```

As you can see, a pointer to a class works much like a pointer to a structure. The use of class pointers can be powerful because it allows you to pass an entire class, with all its data and methods, into a function. And because you are using a pointer, you don't need to worry about returning all those values.

ABSTRACT CLASSES AND PURE VIRTUAL FUNCTIONS

Before we can delve into the topic of abstract classes, we must first explore what a virtual function is. A virtual function is a function that, if inherited, must be overridden.

Let us examine exactly what this means. In a class, you may have one or more functions. If you inherit from that class you have the choice of either using the inherited function as it exists in the base class or over-riding it.

HINT!

You will see the terms *override* and *overwrite* used in different textbooks. They both refer to the same operation. That is why this text purposely uses both terms.

You create virtual functions by adding the word `virtual` to their declaration as shown in the following example.

```
virtual void funca();
```

With virtual functions, if you simply create an instance of the class they are in, then the function is treated as any other normal function. However, **if** the class is inherited, then the virtual function **must** be overwritten. The following code segment illustrates this.

```
class someclass
{
    public:
virtual  void funca();
};
void someclass::funca()
{
  cout << "This is my function in the base class\n";
}
class anotherclass
{
  public:
void funca();
}
void anotherclass::funca()
{
    cout << "This is in my derived class \n";
}
int main()
{
  someclass a;
  anotherclass b;

  a.funca();
   b.funca();
   return 0;
}// end of main
```

As you can see, if you create an instance of this class, then you can use the function as its written; if you inherit from this class, then the derived class will have to write its own implementation. If you don't override the function in the derived class, you will get a compiler error. Another way of looking at virtual functions is to realize that they have no effect on the class they are in. They only affect classes that inherit from that class. If the base class that contains the virtual function is instantiated, then the virtual function behaves just like any other ordinary function. However, if some class inherits from a base class that contains a virtual function, then the derived class will have to override the virtual function.

An *abstract class* is a class defining an interface only; used as a base class. An abstract class cannot be instantiated. It can only be inherited from. Declaring a member function of a class as a pure virtual function makes the class abstract and prevents creation of objects of the abstract class. Another way to say this is to state that an abstract class is a class that cannot be instantiated, but must be inherited from. An abstract class is created by making any of its member functions a pure virtual function. The way you create a pure virtual function is simply by adding an =0 at the end of the function.

```
virtual void funca() = 0;
```

The *how* of creating an abstract class is actually rather simple. It is the *why* that is problematic. The reason for creating an abstract class is so that you can have a base class that, although it is not appropriate to be used directly, can be inherited. You create an abstract class by making one of its functions a pure virtual function. Remember that a pure virtual function is created by adding =0 to the end of a virtual function's prototype as you see in the following example.

```
class abstract
{
public:
      virtual void pure_v_func() = 0;
};
```

Example 12.6

Step 1: Enter this code into your favorite text editor and save it as *12_05.cpp*.

```
#include <iostream>
using namespace std;
class abstract
{
public:
      virtual void pure_v_func() = 0;
};
class derived: public abstract
{
 public:
        void pure_v_func();
};
void derived::pure_v_func()
{
  cout << "See how this works";
}
int main ()
{
 derived myclass;
 myclass.pure_v_func();

 return 0;
}
```

You see how we now have an abstract base class. This is due to the fact that it contains a pure virtual function. This means you now cannot directly instantiate the class. It also means that all the classes methods must be overwritten in the derived class. The purpose of the abstract class is to force the derived class to have a given interface. Other programming languages, such as *Java,* use an object type called *interface* to accomplish this. In C++, the abstract class does the same thing. You can think of an abstract class as a template that determines the specific interface that its derived classes will have, while leaving the implementation up to the derived class.

SUMMARY

This chapter has introduced you to the basics of polymorphism and multiple inheritance. This material, coupled with Chapters 9 and 10, should give you a solid understanding of the basics of object-oriented programming in C++. You will see many of these object-oriented techniques used again and again. Multiple inheritance can be confusing for beginning programmers, so be sure to carefully study this chapter, and work through all the examples. You might even wish to do a few different programs using the same techniques you saw here.

This chapter also tackled some rather advanced topics. Virtual functions, pure virtual functions, and abstract classes are definitely advanced topics. If, as a beginner, your understanding of these topics is a little cloudy, don't worry. If you did fully understand those concepts, then you should congratulate yourself for mastering a rather difficult concept.

REVIEW QUESTIONS

1. What is polymorphism?
2. What does it mean to override a function?
3. How is overriding different from overloading?
4. What is the syntax for a child class to inherit from two base classes (`classa` and `classb`)?
5. What is a function's interface?
6. What is a function's implementation?
7. How do you handle the situation where two base classes have the same function?
8. What is a virtual function?
9. What is an abstract class?
10. How do you create an abstract class?

Advanced Topics in C++

This section will introduce you to some more advanced topics in C++. Chapter 13 will show you the basics of data structures and algorithms. These topics are a cornerstone of computer science and programming. Then, Chapter 14 will show you the basics of games programming. Games programming is introduced for two reasons. The first being that games programming is quite challenging and can stretch your skills as a programmer. The second reason is that you will have worked very hard in the preceding thirteen chapters and it will be time to let our hair down and have some fun!

13

Basic Data Structures and Algorithms

IN THIS CHAPTER
••••••••••••••

- Data Structures
- Algorithms
- The Algorithm Header File
- Recursion
- The Fibonacci Sequence

The first 12 chapters of this book have focused on acquainting you with the language of C++. You have explored the syntax, structure, and techniques of this programming language. However, you should be aware that this is only part of the process of programming. Simply understanding language syntax is not enough. One of the fundamental cornerstones of computer science is the study of data structures and algorithms. In this chapter, you will be introduced to both of these topics.

DATA STRUCTURES

The first order of business is to define exactly what a data structure is. A data structure is an organized way of storing data that also defines how the data will be processed. Structured ways of storing data are quite commonly used—you have already encountered them without knowing it. For example, when you send documents to a networked printer, you are sending them to a *queue*, a data structure. A queue is, in fact, one of the most commonly encountered data structures. Let's start with examining queues in more detail.

In a queue, data is entered in sequence from the beginning to the end. The queue is usually represented by an array. The pointer that shows where the last element was added is called the *head*. The pointer that shows where the last element was removed and processed is called the *tail*. Figure 13.1 illustrates this.

When you use a network printer, each element of the queue array is a document (or, more likely, a fixed amount of memory). The head represents the last place that a document was added to the print queue. After

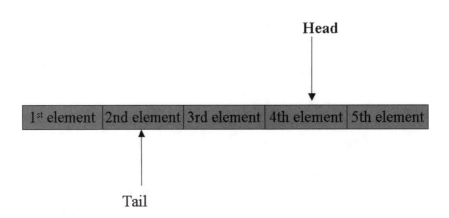

The Queue Data structure

FIGURE 13.1   Structure of a queue.

an item is added, the head moves forward one space. The tail represents the processing end of the queue. Once an item is processed (i.e., printed) it is removed from the queue and the tail is advanced. When the head reaches the end of the queue, it starts back over at the beginning, and that is what is referred to as a *circular queue*. Let's look at an example using a simple queue of integers.

Example 13.1

Step 1: Enter the following code into your favorite text editor.

```
#include <iostream>
using namespace std;
int queue[10];
// both the head and the tail will start
// out pointing to the first element in
// the array
int *phead = &queue[0];
int *ptail = &queue[0];
int main( )
{
    int i;
    for( i= 0;i<10;i++)
    {
    cout << "Please enter an integer. \n";
    cin >> queue[i];
    // increment the head pointer so that
    // it points to the next element of the array
    phead++;
    }
    for(i=0;i<10;i++)
      {
    ptail++;
    cout << queue[i] << endl;
      }
      return 0;
}
```

Step 2: Compile and execute that code. You should see something much like what is depicted in Figure 13.2.

You should notice that, in this example, the pointer's head and tail are only being used to keep the place of where we are currently. In some

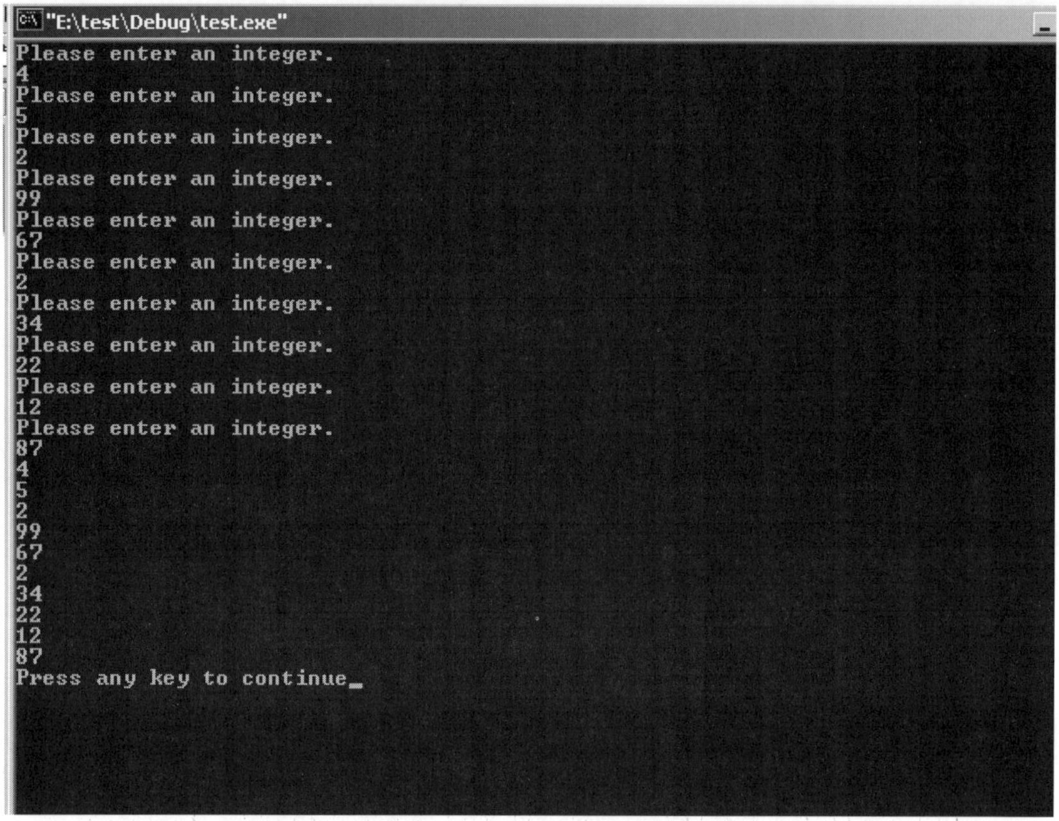

```
"E:\test\Debug\test.exe"

Please enter an integer.
4
Please enter an integer.
5
Please enter an integer.
2
Please enter an integer.
99
Please enter an integer.
67
Please enter an integer.
2
Please enter an integer.
34
Please enter an integer.
22
Please enter an integer.
12
Please enter an integer.
87
4
5
2
99
67
2
34
22
12
87
Press any key to continue_
```

FIGURE 13.2 Circular queue demonstration.

cases, programmers will use a plain integer in this case rather than a pointer to an integer. The real problem occurs when your are processing items out, slower than entering them in. You see if the head catches the tail, it will start overwriting things you have not yet processed! Normally, you would check to see if the head is the same value as the tail. If it is, then you tell the user "queue is full" and you don't take any more input until your processing can catch up.

Another popular data structure is the stack. Whereas a queue processes items on a first-in, first-out basis (FIFO), the stack does its processing on a last-in, first-out basis (LIFO). A stack gets its name from the way it processes data, much like a stack of plates. If you stack up 10 plates, you will have to remove the last one you placed on the stack before you remove any others (unless, of course, you are a magician)). Figure 13.3 illustrates the way a stack works.

The Stack Data structure

Head/Tail work in conjunction. You always remove the one you just added first.

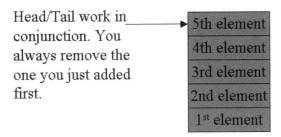

FIGURE 13.3   The stack.

The registers on your computer's central processing unit (CPU) use stacks. Items are pushed onto the stack and popped off. If you worked with an *Assembly* programming language, you would become intimately familiar with the stack. However, for our purposes, it is enough that you know what a stack is.

The stack and the queue are just two examples of data structures. There are many more to choose from. It is beyond the scope of this book to examine all these data structures. However, you should be at least aware of what some of these data structures are. You can then use some of the resources listed in the back of this book to learn more if you so desire. In addition to stacks and queues, you may wish to acquaint yourself with the linked list, double linked list, and binary tree.

One more commonly encountered data structure is the *linked list*. A linked list consists of a series of nodes (made either with structures or classes) that have linked to the next node in the series. The link to the next node is usually accomplished with a pointer to another node. The following is an example of a node.

```
struct node
{
 int data; // data can be of any type at all.
 node *nextnode;
}
```

Each node can then be included in a class. The class must have at least two methods. The first method is a push method to add new nodes to the linked list. The second method is a pop to remove nodes from the linked list. Because of the push and pop methods, a linked list is capable of dynamic sizing. That means that it can grow larger or smaller. This is in direct contrast to an array, which cannot change its size. The next step is to build a class to house the nodes, as well as the push and pop functions. The following example shows you how to create and use a linked list.

Example 13.2

Step 1: Place the following code into your favorite text editor and save it as *linkedlist.h.*

```
struct node
{
 int data; // data can be of any type at all.
 node *nextnode;
      node *prevnode;
};
class linkedlist
{
private:
  node base;    // This is the base node
     // from which other nodes will
// spring.
  node *curnode;// A pointer to the current node.
public:
    void push(int item);
    int pop();
};
void linkedlist::push(int item)
{
    // This function basically creates a new node and
    // places it at the top of the linked list. The node
    // currently at the top is moved down.
  node *temp=curnode;         // Temporary node
  curnode->nextnode=new node;// Create new node
  curnode=curnode->nextnode; // Assign current to
                              // new top node
```

```
 curnode->data=item;    // Assign the data
 curnode->prevnode=temp;// Set previous
   // pointer
}
int linkedlist::pop()
{
    // This function pulls the data off the current
 // node,then
    // moves the node preceding the current node into the
 // front
    // of the linked list.
 int temp;
 temp=curnode->data;
    // if the previous node is somehow already gone, then
 // skip
    // the rest.
 if (curnode->prevnode == NULL)
   return temp;
  curnode=curnode->prevnode;
     // delete the node we just popped off
  delete curnode->nextnode;
  return temp;
}
```

Step 2: Write the following code into your favorite text editor and save it as *13_02.cpp*.

```cpp
#include "linkedlist.h"
#include <iostream>
using namespace std;
    int main()
{
    linkedlist mylist;
    int data;
    cout << "Enter your data, and enter -1 to exit \n";
    while(data != -1)
    {
 cout << "Enter your next number \n";
      cin>> data;
      mylist.push(data);
    }
    return 0;
}// end main
```

This data structure undoubtedly appears much more complicated than the queue. There are extensive comments in the source code to aid in your understanding. However, let's take a moment to review the highlights. To begin with, the core of the linked list is the node structure. It holds the data. Now, in our case, the data is an integer; however, it could be any data type at all, including another structure or class.

This structure is encapsulated in a class. We have a pointer that tells us the current node and the next node. We then have two functions that allow us to add new nodes or remove old ones.

HINT!

 A doubly linked list is a linked list with pointers to the next node and the previous node.

ALGORITHMS

Computer programs are used to solve problems; that is the entire reason people write computer programs. Unfortunately, many beginning computer programmers approach computer programming in an ad hoc manner, simply cobbling together a patchwork of code that hopefully will properly process their data. There is a much more efficient way to solve problems. This method is called an algorithm. An algorithm is a formal methodology for solving a problem. It is a step-by-step recipe for performing some action. When you take a mathematics course in high school or college, you are taught a number of algorithms—methods for solving problems. In math, these problems involve finding the slope of an equation, finding the length of one side of a triangle, or using integral calculus to calculate the area of function. Another way to phrase that definition of an algorithm is to say that an algorithm is a precise definition of a computation.

Computer scientists study algorithms and the efficacy of algorithms. In a standard baccalaureate computer science degree, students spend an entire semester studying and analyzing algorithms. In graduate-level computer science programs, a deeper study of algorithms is often pursued. That level of algorithm study is well beyond the scope of this book, however we will spend some time studying some of the most common algorithms.

One of the most common algorithms is the sorting algorithm. The need to sort a list is a very common problem. For example, sorting a list of phone numbers into either ascending or descending order is often necessary. Due to the commonality of this problem, computer scientists spend a great deal of time studying sorting algorithms. For our purposes, we will examine the two most common algorithms. These are the *bubble sort* and the *quick sort*.

Bubble Sort

The bubble sort is a simple but effective method for sorting a list of numbers. The bubble sort starts at the beginning of the list. Each element is compared to the element next to it. If the element on the left is larger than the element on the right, then the two numbers are switched. The process is continued until all numbers are in order. It might be helpful to examine an actual code example. In this example, you will see a larger than usual number of comments. It is hoped that the additional commenting will help you to better understand the example.

Example 13.3

Step 1: Enter the following code into a text editor.

```
#include <iostream>
using namespace std;
int main()
{
 int numbers[10],temp,i,size = 10;
 int original,sorted;
 //Creates an array of 10 user supplied numbers.
 for(i=0;i<size;i++)
 {
   cout<<"Please enter a number.\n";
   cin>>numbers[i];
 } // end of for loop
 //outer loop goes through the entire list for(original=1;
original < size; original++)
 {
   for(sorted = size-1; sorted >= original; sorted-)
   {
 /*If element sorted-1 is greater than element after
 it then swap it.*/
```

```
    if(numbers[sorted-1] > numbers[sorted])
      {
         temp = numbers[sorted-1];
    numbers[sorted-1] = numbers[sorted];
    numbers[sorted] = temp;
      }// end of if
        }// end of inner for loop
}// end of outer for loop
//print the sorted array.
cout << "The sorted array is \n";
for(i=0;i<size;i++)
{
 cout<<numbers[i]<<" "<<endl;
}//end of the for loop

return 0;
}// end of main
```

Step 2: Compile the code and execute it. You should see something much like what is displayed in Figure 13.4.

Notice that the first portion of the program is getting the user to enter in 10 integers. The sorting portion is done by a nested loop structure. This is something you may not have previously encountered. A *nested loop* is a loop that is inside another loop. Each time the outer loop executes, the inner loop will go through its complete cycle. This means that this algorithm has to execute a number of times equal to the maximum of the outer loop, times the maximum of the inner loop. This is important because when one studies algorithms, a great deal of attention is devoted to which algorithm is most effective, which one can execute the quickest.

The second loop in the bubble sort holds the key to this algorithm.

```
    if(numbers[sorted-1] > numbers[sorted])
```

The statement simply says that if the number in element i-1 is greater than the number located in element i, then swap them. So if the third element of the array is 5 and the fourth element of the array is 2, then swap them to put them in order. If you wanted to sort the array into descending array instead, then change the *greater than* in the if statement to a *less than*.

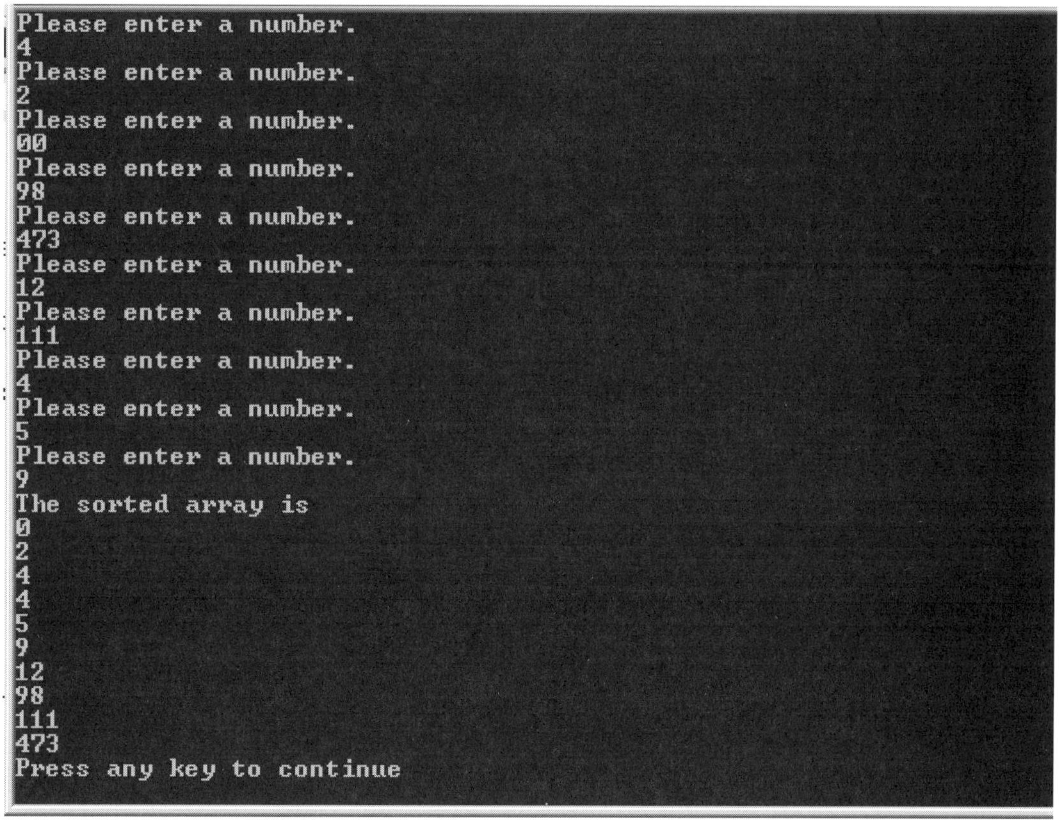

```
Please enter a number.
4
Please enter a number.
2
Please enter a number.
00
Please enter a number.
98
Please enter a number.
473
Please enter a number.
12
Please enter a number.
111
Please enter a number.
4
Please enter a number.
5
Please enter a number.
9
The sorted array is
0
2
4
4
5
9
12
98
111
473
Press any key to continue
```

FIGURE 13.4   The bubble sort.

Quick Sort

Bubble sort and quick sort are the two most commonly encountered sorting algorithms. That's not to say that they are the best sorting algorithms, but that they are the most commonly encountered. The quick sort is based on the idea of splitting an array down the middle and sorting both sides, then putting them back together. If both sections are sorted simultaneously then this should, theoretically, sort faster. The following example is commented rather heavily in an attempt to facilitate your understanding of the code presented. There will also be some commentary after the example.

Example 13.4

Step 1: Enter the following code into your favorite text editor.

```cpp
#include <iostream>
using namespace std;
// function prototypes
int *swapnumbers(int rawdata[], int lower, int upper);
int *quicksort(int rawdata[], int first, int last);
int main()
{
    int unsorted[10],i;
    int *sorted;// pointer to a sorted array
    // enter numbers in array in
    // any order at all.
    for(i = 0;i<10;i++)
    {
    cout << "Please enter an integer \n";
      cin >> unsorted[i];
    }
    // call the quick sort function.
    // Pass it the unsorted array and the starting
    // number of the array.  Also pass it the size of
    // the array.
    sorted = quicksort(unsorted, 0, 10);
    // print out the sorted array
    cout << "This is your array sorted:"<< endl;
for(i = 0; i < 10; i++)
{
    cout << sorted[i] << endl;
}
    return 0;
}
int *quicksort(int rawdata[], int first, int last)
{
    int lower = first+1, upper = last;
    int bound;
    // begin by swapping numbers
    // the first+last /2 gives you the middle
    // point of the array...often called the
    // pivot point
    rawdata = swapnumbers(rawdata, first, (first+last)/2);
    bound = rawdata[first];
```

```
    // break the array into two partitions
    while(lower <= upper)
    {
    while(rawdata[lower] < bound)
    lower++;
    while(bound < rawdata[upper])
    upper-;
    if(lower < upper)
    {
    rawdata = swapnumbers(rawdata, lower++,
upper-);
    }
    else
    {
    lower++;
    }
    }// end of while loop
    rawdata = swapnumbers(rawdata, upper, first);
    // call quicksort for each sub section of the array
    // recall that when a function calls itself, this is
    // referred to as recursion
    if(first < upper-1)
    {
    rawdata = quicksort(rawdata, first, upper-1);
    }
    if(upper+1 < last)
    {
    rawdata = quicksort(rawdata, upper+1, last);
    }
    return(rawdata);
}
int *swapnumbers(int rawdata[], int lower, int upper)
{
    int temp;
    temp = rawdata[lower];
    rawdata[lower] = rawdata[upper];
    rawdata[upper] = temp;
    return(rawdata);
    }
```

Step 2: Compile and execute the code. You should see something much like what is shown in Figure 13.5.

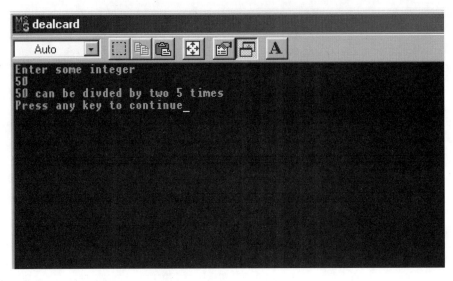

FIGURE 13.5   The quick sort.

This sorting algorithm works a bit differently than the previously demonstrated bubble sort. It basically splits the list and sorts it in sections simultaneously. In many cases, the quick sort will be faster than the bubble sort. However, the bubble sort is perhaps the quickest and easiest sorting algorithm for beginners to understand and implement.

THE ALGORITHM HEADER FILE

The C++ standard library includes an algorithm header file. What you have seen throughout this chapter are algorithms that you wrote yourself. However, many algorithms are already done for you in the algorithm header file. Several of these algorithms are summarized in Table 13.1.

If you include the algorithm header file in your programs, you will have access to these and many other algorithms. It is still a good idea for beginning programmers to code their own sorting algorithms so that they develop a better understanding of how algorithms work.

The list in Table 13.1 is, by no means, exhaustive. It is, rather, meant as a sample of the algorithms already worked out for you. The key point to remember is that many algorithms are already worked out for you and ready for you to implement. You merely need to include the algorithm header file and appropriately call these algorithms.

TABLE 13.1   The Algorithm Header File

Algorithm	Description
partial_sort	A sorting algorithm.
swap	This swaps elements.
sort	Another sorting algorithm.
stable sort	Yet another sorting algorithm.

RECURSION
. . . .

Recursion is a relatively common programming technique. You are bound to encounter it frequently. It is actually a relatively simple concept. Recursion refers to a function that calls itself. This is frequently used in a variety of algorithms, including factoring algorithms. The function continuously calls itself until a final point is reached, such as the number being completely factored, then the function quits calling itself.

Example 13.5

Step 1: Enter the following code into a text editor and save it as *13.05.cpp*.

```
#include <iostream>
using namespace std;
int findmultiplesoftwo(int);
int main()
{
int input,answer;
cout << "Enter some integer \n";
cin >> input;
answer =findmultiplesoftwo(input);
cout << input << " can be divided by two ";
cout << answer << " times \n";
return 0;
}// end of main
int findmultiplesoftwo(int number)
{
// This function finds out how many times
// two will go into whatever number is passed to
// it.  It does this via recursion.
```

```
static int count;
number = number / 2;
count++; //increment the counter
if (number > 2)
{
        findmultiplesoftwo(number);
}
else
{
        return count;
}
    }
```

Step 2: Compile and execute the code. You should see something like what is shown in Figure 13.6.

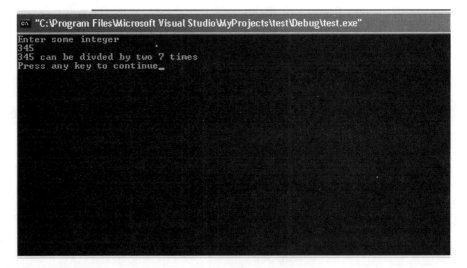

FIGURE 13.6   Recursion.

You can see here that this function calls itself repeatedly. Recursion is often used in more advanced sorting algorithms. If you continue your studies of algorithms, you will undoubtedly encounter recursion. You also saw a new C++ command used here. It was the word static. When a variable is defined locally, within a function, as soon as you exit the function, or call it again, that variable gets recreated. Its value is not retained between iterations of a function. If you put the word `static` before the variable declaration, you are instructing the compiler to retain that variable's value between iterations of the function.

Thus, a recursive function is one that calls itself. A function that does this is said to be recursive. It is important that you be familiar with these two terms, as you will probably encounter them frequently in various programming examples.

THE FIBONACCI SEQUENCE

The *Fibonacci sequence* is an interesting sequence of numbers. It is of particular interest to mathematicians, and is frequently used in computer science to provide a demonstration of recursion. Before we delve into this example of recursion, we should first explore what a Fibonacci number is.

This sequence of numbers was discovered by a mathematician named Fibonacci, thus the name of the sequence. The algorithm used to derive the sequence is quite simple. Examine this series of the first 10 Fibonacci numbers and see if you can derive the algorithm used to produce them.

        1   1   2   3   5   8   13   21   34   55

Can you guess what is being done? Look closely. If you add the first two numbers, a 1 and a 1, you get the third number, a 2. If you then add the second and third numbers, 1 and 2, you get the fourth number, 3. If you add the third and fourth numbers, 2 and 3, you get the fifth number, 5. This sequence continues infinitely.

The exciting part is where this algorithm pops up. This algorithm is seen throughout nature. The seeds of a sunflower, nodes on the shellfish, a rabbit breeding, and the like all exhibit the Fibonacci sequence. This is why this is so interesting to mathematicians. The sequence can be derived, to any length desired, by a simple recursive function. This is why it is of such interest to computer scientists.

Example 13.6

Step 1: Enter the following code into a text editor and save it as *13-06.cpp*.

```
#include <iostream>
using namespace std;
void fibonacci(int, int);
int previous = 1;
int main()
{
int stop;
cout << "Where would you like the fibonacci ";
cout << "sequence computed to \n";
cin >> stop;
cout << endl<<endl;
cout << "Your Fibonacci sequence : "<<endl;
fibonacci(1,stop);
return 0;
   }// end of main
   void fibonacci(int curnum,int max)
   {
int newnum;
newnum = curnum + previous;
previous = curnum;
cout << newnum << endl;
if(newnum <= max)
{
 fibonacci(newnum,max);
}
else
{
 return ;
}
  }
```

Step 2: Compile and execute the code. You should see something much like what is depicted in Figure 13.7.

This example shows you another application for recursive functions. If you carefully examine the code, you see that the user sets the number that will be the cut-off point. That number and the current number being computed are passed to the Fibonacci function. As long as the cut-off has

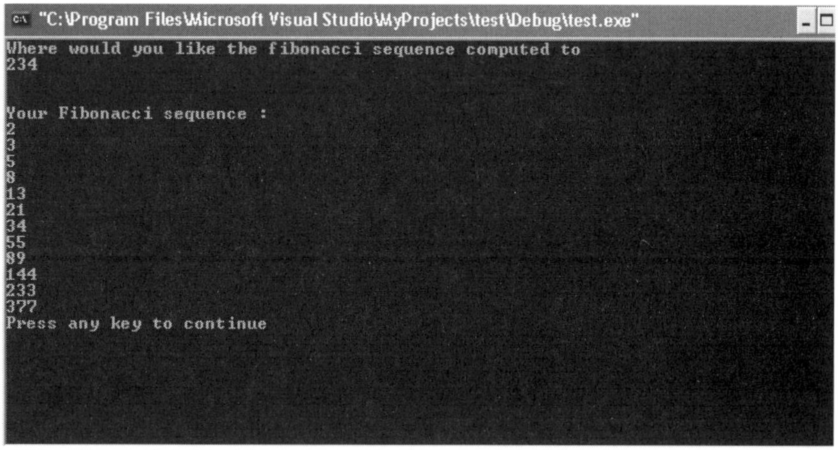

FIGURE 13.7   The Fibonacci sequence.

not been reached, the Fibonacci function will continue to call itself repeatedly, each time printing out the current number in the sequence. You can use the program you just saw in the previous example to compute the Fibonacci sequence as far out as you like.

SUMMARY

This chapter has given you an introduction to an important concept in computer science, data structures and algorithms. You have seen how to write a linked list and a bubble sort. The linked list is a very commonly encountered data structure, and the bubble sort is, perhaps, the most commonly encountered sorting algorithm. A careful study of this chapter gives you an introduction to this field of study. Because sorting algorithms are so commonly encountered, it would be of benefit to you to familiarize yourself with the sorting algorithms demonstrated.

This chapter also introduced you to the concept of recursion. This is a simple, but powerful programming concept that is used in a number of situations, including sorting algorithms. You where also shown the Fibonacci sequence, which is a widely studied mathematical curiosity, used frequently in computer science to implement recursive functions.

REVIEW QUESTIONS

1. What is a data structure?
2. Give three examples of data structures.
3. What is a stack?
4. What is an algorithm?
5. What is the basis of the bubble sort?
6. What two methods must be associated with a linked list?
7. What is a doubly linked list?
8. What does the keyword *static* do?
9. What is recursion?

Build Your Own
Game in C++

IN THIS CHAPTER
....

- The Essentials of Games Programming
- Simple Command-Line Games
- Graphics Games Programming
- Card-Dealing Engine

Games programming can be the most exciting, and the most challenging area in programming. Unlike business programming, where you carefully guide the user through a narrow range of choices, in games programming you want to give the user the most freedom possible. This makes games programming especially difficult. However, learning to write your own games is an excellent learning activity, and it's also a lot of fun!

Games programming is a rather extensive topic. There have been rather large tomes written on the topic. A single chapter is unlikely to make you a game-programming guru. The purpose of this chapter is to introduce you to the concepts of games programming and to show you a few simple command-line-based games. The end of this chapter briefly discusses some ideas behind graphics games programming, and *Windows*-based games programming. You will, however, see a few complete command-line-based games. If you are interested in games programming, this will hopefully encourage you to explore this area in more depth.

THE ESSENTIALS OF GAMES PROGRAMMING

Games programming is not only a fascinating arena of programming but also challenging and educational. Many of the concepts introduced throughout this book such as loops, nested if statements, classes, algorithms, and more are used in games programming. Certain concepts that were introduced earlier in this book will take on more significance now. Loops are one such example. It is virtually impossible to write a game without some sort of looping structure. In a game, you want the code to continue executing until the user wishes to end the game. The usual way to do this is to have some loop surround the primary function of the game and to keep calling it until such time as the user enters some key to stop the game:

```
while (!stopgame)
{
    startgame();
}
```

You can see that loops can be quite essential to games programming. They are key to keeping the game operating until the user is done playing. Loops are also often used to draw graphics. Loops can be quite useful in any graphics programming, but they are absolutely critical for command-line-based graphics. When you are trying to use basic cout statements to display simple graphics, loops are almost the only way to accomplish this. In the example of a tic-tac-toe game that you will see later in this chapter, for loops are used to draw the three rows of the Tic Tac Toe board.

Another familiar programming technique, that is of vital importance to games programming, is the if statement. Clearly any game, even a very simplistic one, must take different courses of action depending on what decisions the user makes. if statements are critical for branching action, depending on the user's choices. Of course, with more complex choices a switch statement will be needed; however, the premise is the same. Responding to user input in an interactive manner is useful in any field of programming; it is especially critical to games programming.

Classes are also frequently used in all types of games programming. In simpler games, you may need an object to represent a deck of cards, a tic-tac-toe board, or dice. With more complex, graphics-oriented games, classes play an even bigger role. For example, in a flight simulator, you probably have a class that represents the airplane, the parts of the plane, other objects in the sky, and so on.

The real essence of games programming is to take the standard programming techniques you have already learned in this book and apply them in new and creative ways. The real secret to games programming is creative thinking.

SIMPLE COMMAND-LINE GAMES

Rather than discuss the various aspects of a simple game in abstract terms, it would probably be useful for us to take a look at the code for a simple game. Example 14.1 shows you a fully working, two-player Tic Tac Toe game. Before you dive into the code, it would probably be a good idea to have a basic idea of what each of the functions in the code do. Table 14.1 summarizes each function and its purpose.

TABLE 14.1   The Functions and Variables of the Tic Tac Toe Game

Function/Variables	Purpose
StartGame()	This function simply starts the game.
Next_Player_Moves()	This function is used after one player moves.
Move()	This function does the initial move for the game.
Check_For_Win()	This function checks the various possible winning combinations, to see if any player has won.
DrawBoard()	This function draws the board on the screen.
Board	This array is a 3 by 3 array of integers that represent the actual Tic Tac Toe board.
turn	This contains a number indicating whether player one or player two is the current player.
IsGameOver	This is an integer value representing whether the game has been won.

In addition to this table, there are extensive comments throughout the code, explaining each function and variable. This should make the code relatively easy to follow and understand. It is also important for the reader to note that the actual code consists of basic techniques that have already been covered in depth in this book. The real key to games programming comes from the creative use of these functions.

ON THE CD HINT!

Some of the lines of code in Example 14.1 are a bit long and may wrap to the next line. If you have trouble following those portions of code, just look at the example on the accompanying CD-ROM.

Example 14.1

Step 1: Enter the following code into your favorite text editor and save it as *14-01.h*.

```cpp
#include <iostream>
using namespace std;
// constants to show whose turn ti is
const int PLAYER1 = 1;
const int PLAYER2 =2;
class TicTacToe
{
 int board[3][3];  // 3 X 3 board to play on
 int turn;// whose turn is it
 int isGameOver;   // is the game over?
 void PlayGame();  // start game
public:
    // functions for making
    // moves and starting the game
    void StartGame();
    void Next_Player_Moves();
 int Move(int);
    // other functions
 void Check_For_Win();
 void DrawBoard();
};
void TicTacToe::StartGame()
```

```
{
     // This function clears the board and starts the game
for(int i = 0; i < 3; i++)
for(int j = 0; j < 3; j++)
 board[i][j] = 0;// A zero is indicative of an empty
     // square
 turn = PLAYER1;
 isGameOver = 0;
 DrawBoard();  // This function draws the game
 PlayGame();    // This starts playing the game
}
int TicTacToe::Move(int i)
{
     // the math is all based on three's
     // since each row has three choices.
 int x = (i - 1)/3;
 int y = ((i + 2) % 3);
 int returnVal = board[x][y];
 if (returnVal == 0)
    {
   board[x][y] = turn;
   Check_For_Win();
   if (!isGameOver)
      {
 Next_Player_Moves();
      }// end of inner if
      }// end of outer if
  else
  cout << "Invalid move, try again.\n";
     DrawBoard();// Redraw the board
     return returnVal;
}
void TicTacToe::Next_Player_Moves()
{
     // This function checks to see whose turn it was last
     // then gives the other player a turn
   if (turn == 1)
 turn = PLAYER2;
   else
 turn = PLAYER1;
}
void TicTacToe::Check_For_Win()
{
     // This checks to see if the game is over, and who
```

```
      // has won it. It does this by checking the various
      // possible combinations that might lead to a win. This
      // is done in a series of nested if statements checking
      // each possible win scenario.
 if ((board[0][0] == turn) && (board[1][0] == turn) &&
(board[2][0] == turn))
   {isGameOver = turn;}
 else
   if ((board[0][1] == turn) && (board[1][1] == turn)
&& (board[2][1] == turn))
 {isGameOver = turn;}
   else
if ((board[0][2] == turn) && (board[1][2] ==
  turn) && (board[2][2] == turn))
   {isGameOver = turn;}
else
   if ((board[0][0] == turn) && (board[0][1] ==
turn) && (board[0][2] == turn))
{isGameOver = turn;}
   else
if ((board[1][0] == turn) && (board[1][1] ==
  turn) && (board[1][2] == turn))
   {isGameOver = turn;}
else
   if ((board[2][0] == turn) && (board[2][1]
== turn) && (board[2][2] == turn))
{isGameOver = turn;}
   else
if ((board[0][0] == turn) &&
  (board[1][1] == turn) && (board[2][2]
== turn))
   {isGameOver = turn;}
else
   if ((board[0][2] == turn) &&
(board[1][1] == turn) && (board[2][0] == turn))
  isGameOver = turn;
}
void TicTacToe::PlayGame()
{
  int imove;
     // if the game is not over (i.e. if isGameOver!= turn)
     // then prompt for the next move.
while (isGameOver!=turn)
     {
```

```
    //DrawBoard();
     cout << "Player Number " << turn << " Please enter
    move: \n";
     cin >> imove;
     Move(imove);
}
        // when a player wins the while loop will stop
        // and we can display the winner
cout << "Player Number " << turn << "  Wins!" << endl;
}
void TicTacToe::DrawBoard()
{
        // this function simply draws the
        // Tic Tac Toe board.
        int temp[9];
int k = 0;
for(int i = 0; i < 3; i++)
for(int j = 0; j < 3; j++)
    {
    if (board[i][j] == 0)
  temp[k] = k+49;
      else
      {
   if (board[i][j] == 1)
temp[k] = 88;    // this creates an x
   else
temp[k] = 79;    // this creates a o
    }
    k++;// end of outer if
}       // end of inner for loop
     cout << "*****2-Player Tic Tac Toe*****\n";
 cout << " |**********|\n";
 cout <<" |" << (char)temp[0] << " | " <<
(char)temp[1] << " | " << (char)temp[2] << " | \n";
cout << " |**********|\n";
cout <<" |" << (char)temp[3] << " | " <<
(char)temp[4] << " | " << (char)temp[5] << " | \n";
cout << " |**********|\n";
cout <<" |" << (char)temp[6] << " | " <<
(char)temp[7] << " | " << (char)temp[8] << " | \n";
cout << endl << endl;
}
```

Step 2: Enter this code into a text editor and save it as *14-01.cpp.*

```cpp
#include "14-01.h"
#include <iostream>
using namespace std;
// prototype the menu function
void menu();
// create an instance of the Tic Tac Toe class
TicTacToe mygame;
int main()
{
    // the menu is in a separate function thus making
    // it easier to repeatedly call it, allowing you to keep
    // playing Tic Tac Toe.

      menu();
    return 0;
}
void menu()
{
    char choice; // this finds out it the user wisher
  // to to continue or not.

cout << "Would you like to play a game of Tic Tac To?"
cout << "Type y for yes n for no\n";
    cin.get(choice);
    if (choice=='y')
{
    mygame.StartGame();
    menu();
    }
    else
    {
      return;
    }
}
```

Step 3: When you run the program, you will see images like those shown in Figures 14.1 and 14.2.

In this game, we used basic code and a rather simple class structure to create an interactive two-player game. The actual flow of this game is fairly straightforward and is depicted in Figure 14.3.

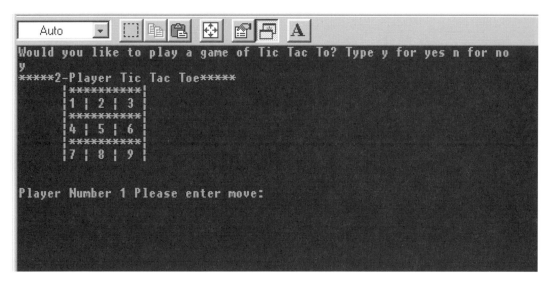

FIGURE 14.1   Intro to the Tic Tac Toe game.

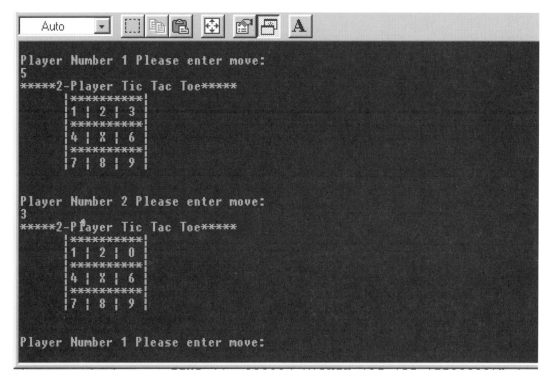

FIGURE 14.2   Playing the Tic Tac Toe game.

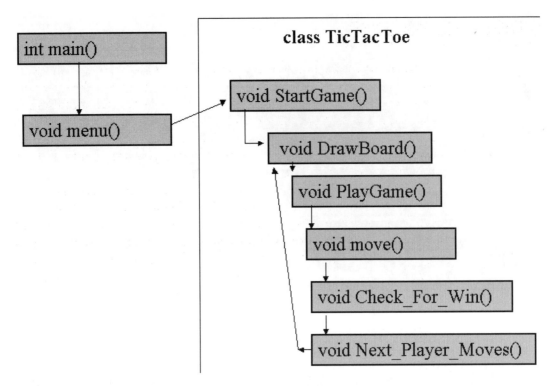

FIGURE 14.3   The flow of the Tic Tac Toe game.

The flow starts with the int main function, as all C++ programs begin. Main then calls the function menu, which will call the StartGame function of the TicTacToe class. The StartGame function is, of course, the starting point for the Tic Tac Toe game. It calls the DrawBoard function that draws the actual 3 by 3 Tic Tac Toe game board, and then calls the move function, that makes the very first move. After that move, the Next_Player_Moves function is called. That move function will continually call the Check_For_Win function. If no one has yet won, then the Next_Player_Moves function is called. This loop is the key to keeping the game moving. Recall that it was mentioned earlier that loops are critical in games programming. And, of course, the Check_For_Win function is based entirely on a series of nested if statements that check to see if any winning condition for the game has been met.

GRAPHICS GAMES PROGRAMMING

While command-line games can be interesting and even quite fun, users have come to expect a lot more from games in our current day and age. Going into any depth into high-end graphics games programming is beyond the scope of an introductory programming text. However, it is possible to give you a brief introduction into this exciting area of programming, and some suggestions as to where you can turn to explore this exciting topic in more depth.

For *Windows* programmers (you will be introduced to *Windows* programming in Chapters 15 and 16), you have an awesome tool at your disposal. That tool is the *Direct X Library*™. This library is a set of classes and functions that you can use to create a wide range of graphics and sounds operations. Essentially, *Direct X* is a multimedia library that is accessible to *Windows* programmers. If you have any experience playing *Windows* games then you probably have installed some game wherein the *Direct X Library* was installed as part of the games installation.

For those of you wanting to delve into graphics on a simpler level, you can try using ASCII graphics. ASCII graphics simply refers to creating graphics with simple ASCII symbols. In other words, simply use the symbols that your keyboard can produce to draw various items. Obviously such drawings will be rather simple, but it is a good place to start. The next example draws an arrow on the screen, using asterisks (*), and then moves the arrow across the screen.

Example 14.2

Step 1: Enter the following code into your favorite text editor and save it as *14-02.cpp*.

```
#include <iostream>
using namespace std;
//function prototypes
void drawarrow(int);
void movearrow();
int main()
{
    for (int j = 0; j<15;)
  {
```

```
        drawarrow(j);
        j +=5;
  }
 cout << endl<<endl<<endl;
 return 0;
    }//end of main
void drawarrow(int offset)
{
 int i;
     // this is used to draw the  top part of
     // the arrow head.
       for (i = 0;i < (8 + offset);i++)
       {
cout << ' ';
       }
       cout << '*';
       cout << endl;
 // this loop shifts the shaft of
     // the arrow, enough spaces to match
     // the offset
     for (i = 0;i<offset;i++)
     {
  cout << ' ';
     }
 // this draws the shaft
     for(i = 0;i<10;i++)
     {
   cout << '*';
 }
     // this moves to a new line
     cout << endl;
     // this is used to draw the  bottom part of
     // the arrow head.
     for (i = 0;i < (8 + offset);i++)
     {
cout << ' ';
     }
     cout << '*';
     cout << endl;

 }// end of drawarrow
```

Step 2: Compile and run the program. You should see an image like the one depicted in Figure 14.4.

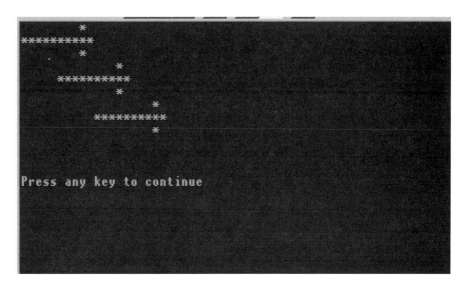

FIGURE 14.4 The asterisk arrow.

This simple graphics illustration illustrates how a little creativity can allow you to create moving graphics from simple keyboard symbols. You also see that loops are used extensively for drawing the arrow.

CARD-DEALING ENGINE

This chapter has shown you how to create a complete working Tic Tac Toe game and how to create basic ASCII graphics. These two items should give you a start on games programming. The complete working example of the Tic Tac Toe game is instructional and should help you understand command-line-based games. However, it is just as important to give you tools that you can apply to your own games. The next example will give you a card-dealing function. The purpose of this function is to deal cards, nothing more. It does not actually play any particular card games. However, if you already have the ability to deal random cards and display them, it is not a particularly arduous task to create a card game.

Example 14.3

Step 1: Enter the following code into your favorite text editor.

```
#include <iostream>
#include <ctime>
using namespace std;
// prototype functions
void dealcards();
int main()
{
dealcards();
return 0;
}// end of main
void dealcards()
{
int cardused[52];   // this array contains the cards
                    // that have already been used.
int cards;  // cards
int face;   // The face of the card selected
int suit;   // The suit of the card selected
char choice;    // deal again or not

// You must seed the random function by
// calling srand and passing it some value.
// Time is the one most commonly used.
srand(time(NULL));
cout << "Here are the cards" <<endl;

for(int i = 0; i < 52; i++)
{

    cards = rand()%52;      // randomize 52 cards
    while (cardused[cards] ==1)
    {
  cards = rand()%52;            //changing the cards
 }
  cardused[cards] = 1;
  face = cards%13;   // calculating the
   // face cards
  suit = cards/13;   // calculating the
   // suit cards

    // the couts use the ascii codes for various symbols
```

```
    // to out put the suit.
    switch(suit)
    {
    case 0:
     cout << '\005';
     break;

    case 1:
     cout << '\006';
     break;

    case 2:
cout << '\x03';
      break;

    case 3:
      cout << '\004';
break;

    }// end of switch

    switch(face)
    {
     case 0:
   cout << "King ";
   break;

case 1:
  cout << "Ace ";
  break;


     case 2:
   cout << "Two ";
   break;

     case 3:
   cout << "Three ";
   break;

      case 4:
   cout << "Four ";
   break;
```

```
      case 5:
   cout << "Five ";
   break;

      case 6:
   cout << "Six ";
   break;

      case 7:
   cout << "Seven ";
   break;

      case 8:
   cout << "Eight ";
   break;

      case 9:
cout << "Nine ";
   break;

      case 10:
   cout << "Ten ";
   break;

      case 11:
   cout << "Jack ";
   break;

      case 12:
   cout << "Queen ";
   break;
    }// end of switch
     cout << endl;

}// for loop
cout << endl << endl;
cout << "Would you like to deal again? Type y for
  yes";

cout << " n for no \n";
cin >> choice;

if (choice == 'y')
    dealcards();
```

```
else
    return;
}// end of the dealcards function
```

Step 2: Compile and run the code. You should see something like what is depicted in Figure 14.5.

FIGURE 14.5 Dealing cards.

You should notice several things about this code. First of all, you should notice that the DealCards function is mostly contained within a for loop. This loop executes 52 times, thus dealing out 52 cards. Obviously for certain card games you would wish to deal 2, 5, 7, or an appropriate number of cards for your game. This is also another example of a game that has a for loop as its most critical component. You should also notice that the first switch statement, the one that picks the face of the card, uses ASCII codes to print-out the symbols for the faces of the cards. It uses these codes to actually print-out a heart, diamond, spade, etc. Again, although this example is not a fully functioning card game, it is the most important component to any card game you may wish to create. Now only your own creative thinking, and, of course, your knowledge of card games, stands between you and the creation of your own computer card games.

SUMMARY

This chapter introduced you to the basics of computer games programming. You were shown how to use standard programming techniques such as loops, if statements, and basic cout statements to create interactive games. We also looked at a few examples of command-line games. These games should give you a basic understanding of how simple games work, and perhaps give you a basis from which to study more advanced games programming.

REVIEW QUESTIONS

1. Why are loop structures so important for games?
2. What is *Direct X?*
3. What is the real key to games programming?
4. Why are if statements so important for games programming?
5. What are some uses of classes in games?
6. What are ASCII-based graphics?

Visual C++

This section will introduce you to the world of *Windows* programming using *Visual C++*. Obviously you will need to have a copy of *Visual C++* to make these examples work. The chapters were written using *Visual C++ 6.0*, but most of the examples will also work with *Visual C++ 5.0*. If you do not have access to either version of *Visual C++*, then you will not be able to use the examples in these chapters.

Visual C++ is designed specifically for *Windows*. This means that although all the ANSI-standard C++ will work in *Visual C++*, there are a LOT of nonstandard items. These items are specific to *Windows* programming, and will only work in *Visual C++*, not in any other compiler.

Clearly this small section will not make you a *Visual C++* guru. However, it should give you a basic working knowledge of *Visual C++*. You can also use some of the resources in the appendices at the back of the book to expand your knowledge of *Visual C++*.

Introduction
to Visual C++

IN THIS CHAPTER

- Windows Architecture
- A Simple Windows Application
- Message Box Function
- Components
- Built-in Functions
- Message Beep Function
- Mouse Events

So far this book has shown you C++ techniques that work with any C++ compiler on any operating system. You could compile the code samples you have seen so far with Borland *C++ Compiler,* Microsoft *Visual C++, GCC++*™, and literally dozens of other compilers. You could also compile and run the code on a *Windows* PC, a *Macintosh,* a *Unix* server, or a *Linux* machine. Standard C++ code is designed to work this way—on any ISO standard C++ compiler with any operating system. C++ is not a platform- specific language. However, there do exist operating-system-specific extensions to the C++ language. Microsoft *Visual C++* is one such extension. The *Visual C++* development tool supports standard C++, but it also includes a lot of *Windows*-specific code. Microsoft *Windows* is a ubiquitous operating system, and its version of C++, *Visual C++*, is in high demand. For this reason, it seems prudent to provide you with a solid introduction to programming with *Visual C++*. For that reason, these last two chapters will be devoted to giving you the basics of *Visual C++*. This will require you to have a copy of *Visual C++*. If you do not, then you should feel free to skip these final chapters.

All of the standard C++ code and techniques you have learned will work just fine in *Visual C++*. There is nothing removed from the standard C++ language. There is, however, a great deal of *Windows*-specific code added. This means that you won't be dismissing any of the skills you have worked so hard to acquire, you will simply be adding to them. It will be important to realize that the *Windows*-specific code will only work with *Visual C++* and will not work with other C++ compilers. This is because it is not ISO standard C++. That is an important distinction to keep in mind. *Visual C++* includes a lot of code that is not ISO standard C++ that will not work on other C++ compilers, or on other operating systems.

WINDOWS ARCHITECTURE

Most people have, at least occasionally, used the *Windows* operating system. However, most casual users, and even many programmers, don't understand how *Windows* works. The first thing to be aware of is that everything you see is a window. The button on a toolbar is a window, the toolbar itself is a window, and the application that the toolbar is in is a window. Each separate component is treated as its own window. Each of these windows must communicate with each other and with the underlying operating system. This communication is handled via messages. If you click on any part of any window, a mouse click message is sent. If you want to do something when such a message is sent, then you will write some function to be executed when that event is received.

These messages are going back and forth all the time. *Windows* has a queue (remember queues?) that the messages are added to. The messages are then processed out and either handled by functions written to respond to specific messages, or simply discarded if there is not handler function assigned. The most important thing for you to remember right now is that everything occurs via these messages.

The second important thing for you to be aware of in the *Windows* architecture, is the Microsoft *Foundation Classes* (MFC). The MFC is a large set of classes built into *Windows*. These classes are designed to handle just about anything you might wish to do. There are classes to make dialog boxes, handle graphics and colors, make buttons, and so on. *Visual C++* is based on the principle of automating much of the use of these foundation classes for you. As we explore *Visual C++*, you will get exposed to many different classes found in the Microsoft *Solutions Framework*.

All standard C++ programs have a main function that returns an `int`. Things are a little bit different with *Windows* programs. Every 32-bit *Windows* application has two primary functions you need to be concerned with: `WinMain` and `WndProc`. Your program will have one `WinMain` for the entire program, and one `WndProc` for each window in the program. `WinMain` is the function that starts your application. Once your application is running, the *Windows* operating system will start placing your applications messages in a message queue. `WinMain` makes three *Windows Application Programming Interface* (API) calls to get these messages from the operating system and to process them. The *Windows* API is simply a large group of functions that are built into the *Windows* operating system. MFC is simply a set of over 200 classes that are wrappers around these API functions.

The messages move from the operating system to your program in a specific order. First, the operating system calls `GetMessage` to retrieve a message. Then, the operating system calls `TranslateMessage` to perform any necessary conversion of the message. Finally, `WinMain` calls `DispatchMessage`, which tells the operating system to send the message to the appropriate `WndProc` for handling.

A SIMPLE WINDOWS APPLICATION

Visual C++ offers several types of projects you can work with. When you first launch *Visual C++* you will be looking at a blank *Visual C++* environment window, much like the one depicted in Figure 15.1.

You then choose `File` from the drop-down menu, then select `New`. You will be given the opportunity to choose one of the many project types available to you in *Visual C++*. This will look similar to Figure 15.2.

As you can see, there are several types of projects you can create with *Visual C++*. Table 15.1 summarizes the most commonly used project types and their purpose.

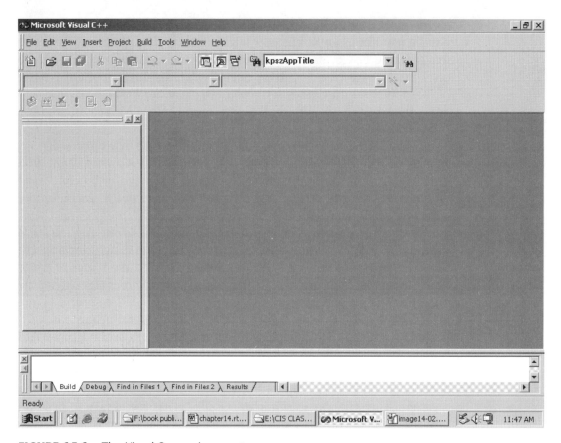

FIGURE 15.1   The *Visual* C++ environment.

The code you have written throughout this book would be done as a *Win32 Console Application*, if using Microsoft *Visual Studio*. The projects in this chapter and the next will mostly be done with *MFC App Wizard (exe)* project types. The *App Wizard* (see Figure 15.3) walks you through, step by step, in creating various types of applications. The wizard is concerned with three types of applications: dialog, MDI, and SDI. Table 15.2 summarizes these types of applications and their purposes.

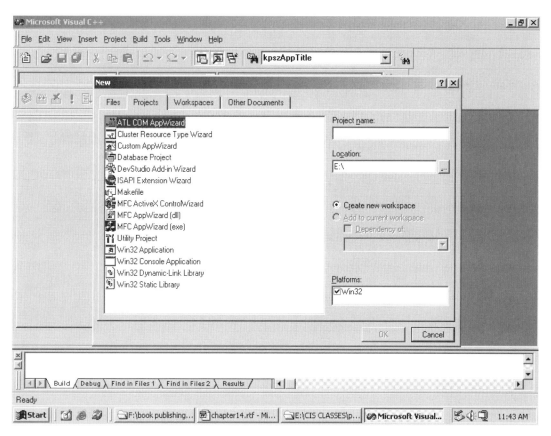

FIGURE 15.2   *Visual C++ project types.*

Our first example will be a dialog application that will illustrate the basics of the *Visual C++ Application wizard, Windows* messaging, and using components in *Visual C++.*

TABLE 15.1   *Visual C++ Project Types*

VC++ Project Type	Purpose
ATL COM Appwizard	ATL is short for *Active Template Library*. It is a quicker way of developing COM components. These are beyond the scope of this book.
Database Project	This one should be relatively easy to understand. It is simply a project that supports connection to a database.
Dev Studio Add-In Wizard	This type of project is for developing add-ins to Microsoft *Developer Studio*...this is also beyond the scope of this book.
ISAPI Extension Wizard	ISAPI is a set of function calls used to create Internet applications.
MFC ActiveX Control Wizard	*Active X* controls are the small components you use in *Windows* applications (buttons, drop-down menus, etc.). This project type is for creating your own *Active X* components.
MFC App Wizard (DLL)	DLLs are *Dynamic Linked Libraries*, or groups of functions with no user interface that are used by other programs.
MFC App Wizard (exe)	This type of project is for standard executables.
Utility Project	A container for files you build without a linking step.
Win32 Application	A non-active X 32-bit *Windows* application.
Win32 Console Application	This project type is for creating standard C++ apps that run in the DOS window/command prompt.
Win32 Dynamic Link Library	These project types are also *DLL* projects, but they don't use *Active X* as their basis.
Win32 Static Library	This project type is for creating static library files with functions.

TABLE 15.2   *App Wizard* Application Types

Application Type	Purpose
Dialog	These are the simplest types of applications and usually consist of a single screen with a few components on it.
SDI (Single Document Interface)	A single document interface allows you to view a single item or document at a time. *Notepad* and *Wordpad* are SDI applications.
MDI (Multiple Document Interface)	A multiple document interface will allow you to view multiple items/documents simultaneously. Microsoft *Word* is an MDI application.

Example 15.1

Step 1: Open *Visual C++*, choose new, and pick the *MFC App Wizard (exe)* type. Name it *15_01*.

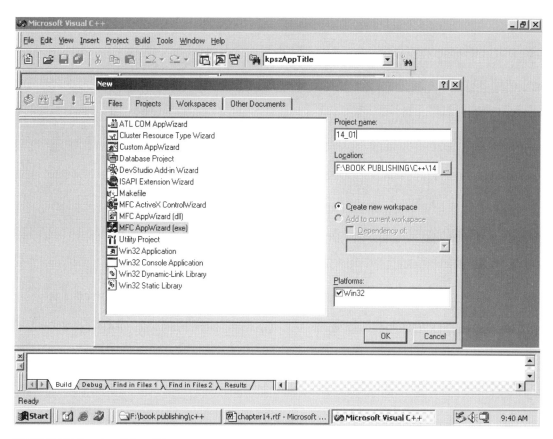

FIGURE 15.3   *App Wizard* first screen.

Step 2: On the next screen (Figure 15.4), choose the dialog application type then press the next button.

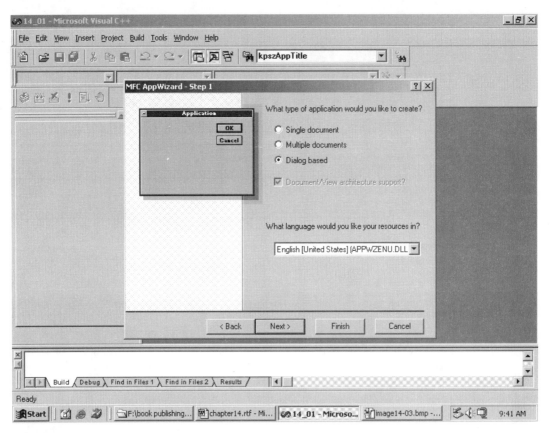

FIGURE 15.4   *App Wizard* second screen.

Step 3: For the third screen (Figure 15.5), simply leave the default selections, then click next.

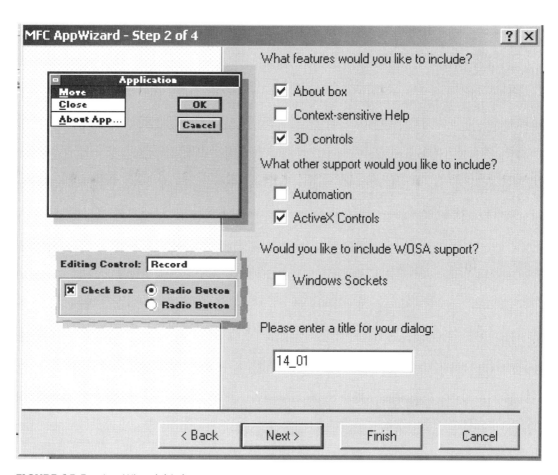

FIGURE 15.5   *App Wizard* third screen.

Step 4: For the fourth screen (Figure 15.6), again use the default values.

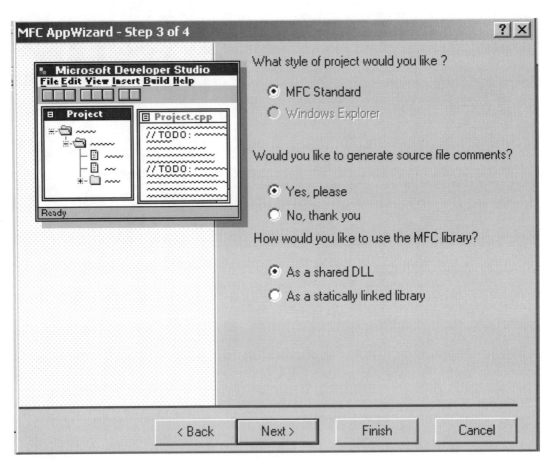

FIGURE 15.6 *App Wizard* fourth screen.

Step 5: For the fifth screen (Figure 15.7), again use the default settings.

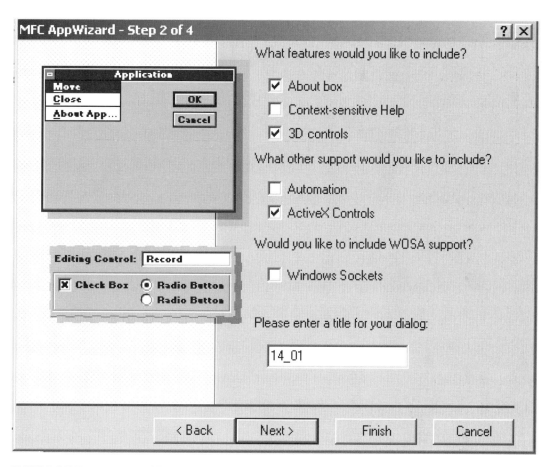

FIGURE 15.7 *App Wizard* final screen.

Now simply click finish and you will be done with the wizard. We will see in just a moment what the wizard has done for you. At this point, you will be looking at a screen much like the one depicted in Figure 15.8.

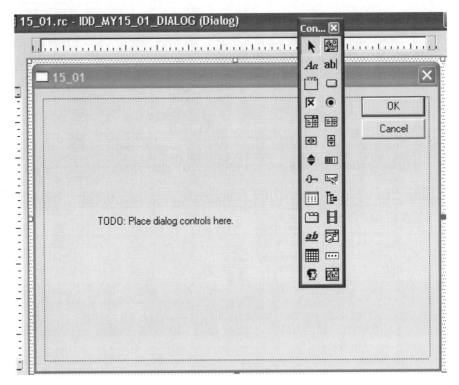

FIGURE 15.8   *App Wizard* sixth screen.

Visual C++'s application wizard has created the basic dialog box for you, with a lot of code (which we will get to in a short while). You should also notice the floating toolbox. In it are several basic components that you can put on your dialog box (or in other programs) to provide basic *Window's* functionality.

Step 6: The basic components in that toolbox will be explained in just a bit. For now, find the Edit box, it should be the one on the top right, and click on it. Then draw an edit box on your dialog box. When you place your mouse over any component in the toolbox, a tag tip will appear telling you what component that is. This is shown in Figure 15.9.

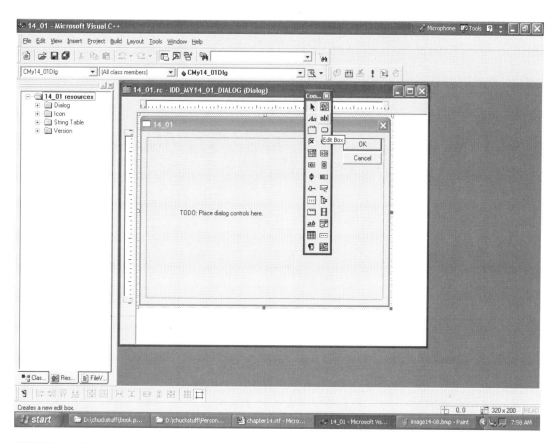

FIGURE 15.9   Toolbox tag tips.

Next find the button component and place one on the dialog box. Then click once on the text that says "TO DO:...." Delete this text. When you are done, your dialog box should look much like the one shown in Figure 15.10.

Now you have finished with the visual components of your dialog box . . . now for the coding part. Note that coding *Windows* is complex; however, there are wizards to do a lot of this work for you.

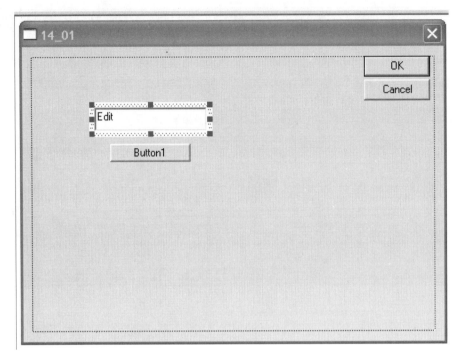

FIGURE 15.10   First dialog application.

Step 7: If you will place your mouse over the Edit box and right click on it you will see a pop-up menu with several choices. Select the *class wizard* and you will be presented with a screen much like the one depicted in Figure 15.11.

HINT!

The *class wizard* is used throughout *Visual* C++ so it is a good idea to get very comfortable using it.

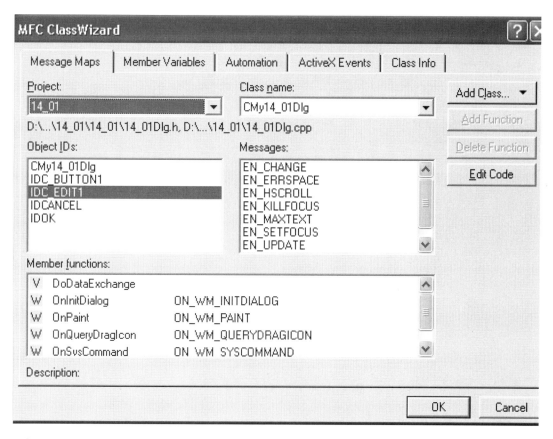

FIGURE 15.11   The *class wizard.*

Step 8: You will need to select the second tab, the one labeled member
variables, and then we will click the add variable button on the righthand side. You will see something much like what is shown in Figure
15.12. Give the variable a name such as mystring or mytext.

HINT!

The m_ preceding variable names is simply to indicate that
these are member variables (i.e., members of some class).

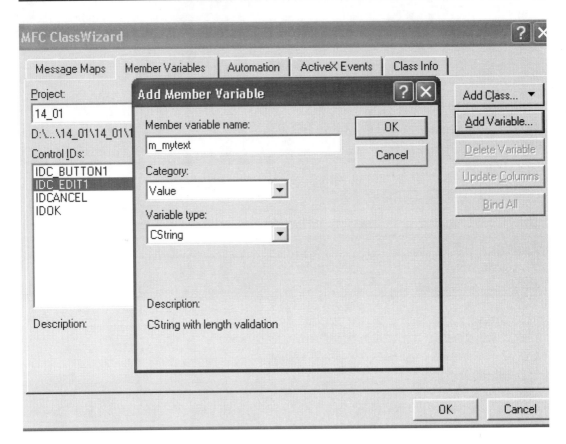

FIGURE 15.12   Adding member variables.

Step 9: You can then click OK to dismiss that window. Next, return to the first tab, the one labeled message maps. Here we are going to create a function that is associated with the button class and which responds to the *Windows* message buttonclick. This means that when someone clicks that button, and *Windows* sends out a buttonclick message, our function will respond to it. When you select the button, you will see an image much like what is shown in Figure 15.13. When you choose add function, you will be taken to a screen that looks like Figure 15.14.

You can then click OK on that window, and then OK on the *class wizard* window. You will be taken back to the main screen.

Step 10: Right-click on the button then select properties. You will be presented with a window where you can change the various properties.

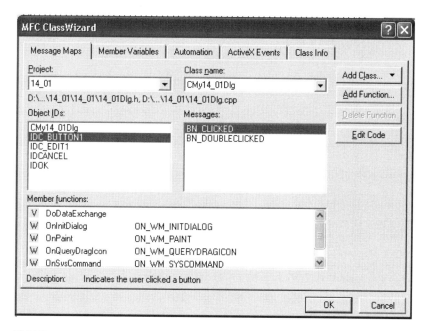

FIGURE 15.13   Selecting a component in *class wizard*.

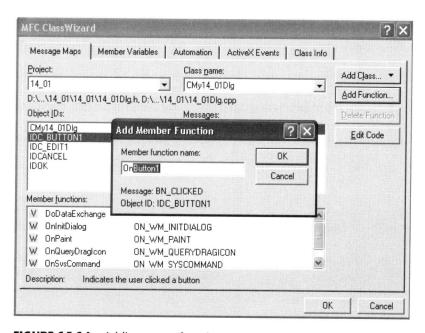

FIGURE 15.14   Adding a new function.

Change the caption to read Click Me. It should look like what you see in Figure 15.15.

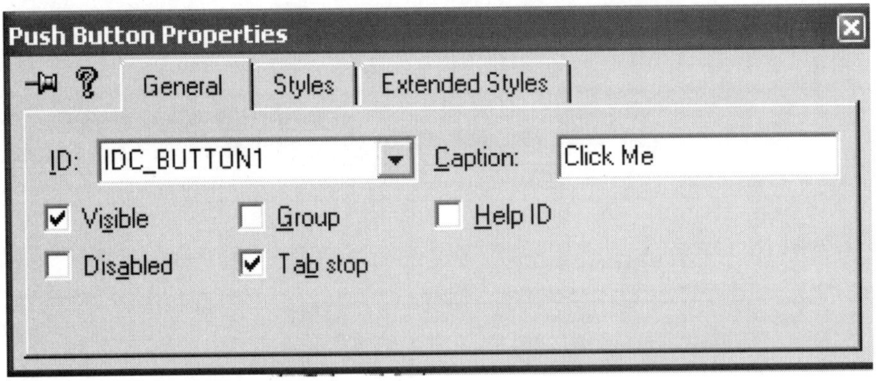

FIGURE 15.15   The properties window.

Step 11: Double-click on the button. It will open to the function we created with the *class wizard*. You will add the following two lines of code.

```
UpdateData(TRUE);
AfxMessageBox(m_mytext);
```

Now run your program. (You can choose Build from the drop-down menu, then select Execute, or you can simply click on the red exclamation mark on the toolbar.) Write anything you wish in the Edit box, then click the button. You should see something like what is shown in Figure 15.16.

Congratulations! You have just written your first *Visual C++* program! And we were able to cover a lot of fundamental material about *Visual C++* programming in the process. However, this leads to a number of questions that we will now explore.

First, we should discuss the *class wizard*. Everything in *Visual C++* is associated with a class. The *class wizard* simply gives you a quick and painless way to associate variables and functions to the classes used for various components (like buttons and edit boxes). We will examine the *class wizard* in detail later in this chapter. For now you should understand that it is essentially a wizard that makes it easy for you to associate components with variables and with functions.

When you associate a component with a variable, that component's contents can be placed in the variable. Also the contents of the variable can be placed in the component. The way you move the contents of the

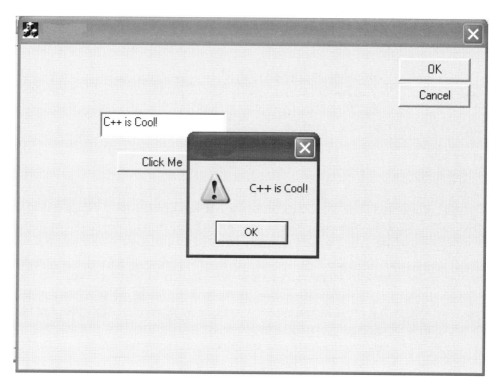

FIGURE 15.16   Running your first dialog application.

variable to the component is to call the method UpdateData(FALSE). This
makes all components update their data to their associated variables. If
you call the method UpdateData(TRUE), then all components get data
from their associated variables. You will see the UpdateData method frequently in the code samples presented in this chapter and the next. The
next item you should notice about the preceding example is the message
box function, which will be explained in the next section.

MESSAGE BOX FUNCTION

In the previous example, we used the message box function to display
data to the user. You have probably seen pop-up message boxes count-
less times in various *Windows* programs. These are such an integral part
of *Windows* programming that it will be vital that you have a good
understanding of them. The first think you must remember is that before

you can utilize the message box function, you must include the following line of code.

```
#include <stdafx.h> .
```

The format of the message box function is simple, and is shown in the following example.

```
AfxMessageBox( text, type, help id)
```

The text is simply whatever text you wish to appear in your message box. The type determines what kind of buttons you want to appear on your message box as well as any icons you may wish to display on the message box. The message box buttons are summarized in Table 15.3.

It is not necessary that you commit all these possible message box settings to memory. You can refer to this table whenever you need to, and you will see several of these settings used in projects throughout this chapter and the next chapter as well. You will, undoubtedly, find the message box to be a very versatile function. It is an excellent way to provide the user with information, and to get their input.

The message box function you saw in the preceding examples simply presented information to the user. The question that we did not address in that example was the following: How do you respond when the user presses a button? How do you even know which button they pressed, assuming you used one of the options to display multiple buttons? When the user clicks on a button, the function returns an integer that tells you which button was pressed. There are a number of constants that will identify the button pressed. Those constants are listed here.

- IDYES—The Yes button was pressed.
- IDNO—The No button was pressed.
- IDABORT—The user pressed the abort button.
- IDCANCEL—The user pressed the cancel button.
- IDIGNORE—The Ignore button was pressed.
- IDOK—The OK button was pressed.
- IDRETRY—The Retry button was pressed.

For example, you could have changed the message box from the previous example, to include responses. The following example illustrates this.

```
int ireturn;
 UpdateData(TRUE);
 ireturn =AfxMessageBox(m_mytext,MB_YESNO);
 if(ireturn == IDYES)
 {
//do something
 }
```

Thus, you can see that you can use the message box function to handle a great deal of user interaction in your *Visual C++* programs.

TABLE 15.3   Message Box Types

Type	Purpose
MB_ABORTRETRYIGNORE	The message box contains: Abort, Retry, and Ignore.
MB_OK	The message box contains just the OK button.
MB_OKCANCEL	The message box contains OK and Cancel.
MB_RETRYCANCEL	The message box contains Retry and Cancel.
MB_YESNO	The message box contains two pushbuttons: Yes and No.
MB_YESNOCANCEL	The message box contains three pushbuttons: Yes, No, and Cancel.
MB_ICONEXCLAMATION	An exclamation-point icon appears in the message box
MB_ICONINFORMATION	An icon consisting of an i in a circle appears in the message box.
MB_ICONQUESTION	A question-mark icon appears in the message box.
MB_ICONSTOP	A stop-sign icon appears in the message box.
MB_APPLMODAL	The user must respond to the message box before continuing work in the current window.
MB_SYSTEMMODAL	All applications are suspended until the user responds to the message box.
MB_DEFBUTTON1	The first button is the default.
MB_DEFBUTTON2	The second button is the default.
MB_DEFBUTTON3	The third button is the default.

COMPONENTS

As was previously mentioned, *Visual C++* offers a number of components in the toolbox that allow you to create standard *Windows* features. Each of these components has different properties and methods and is used in a different manner. These components are summarized in Table 15.4.

There are even more components in *Windows* that you can manipulate inside of *Visual C++*, but these are the basics and should be sufficient for most applications. You have already seen the use of the Edit box and the Button. You will see several of the other components utilized in this chapter and in the next. A few of these components, the ones most commonly used, and the methods for using them will be addressed in this section.

Button

The Button is used to allow the user to initiate action by clicking the button. The most commonly associated function would be OnButton, which is associated with clicking the button. In addition to the obvious properties, such as disabled, the properties tab of the button has a number of settings you may wish to choose. You can choose to have an icon on your button, to have the text vertically aligned, as well as several other options.

Combo Box

This component has a few special caveats you need to be aware of. To begin with, when you draw the component on the form, the area you draw it in, is how far it will drop down when a user selects it. This means that if you only draw it about X inches tall, then it will only drop X inches when selected by the user!

Another important thing to know about the Combo Box is how you add data. You do this by right-clicking on the Combo box to get its properties window, then selecting the second tab, title data. All you have to do then is type in each item you want in the Combo box and press ctl-enter after each entry. You can see this shown clearly in Figure 15.17A.

TABLE 15.4   Basic Windows Components

Component	Purpose
Picture	This component is used simply to display images, such as BMP, JPG, GIF, and the like.
Static Text	This component displays read-only text. The user can view it, but cannot change it.
Edit Box	This component is used to allow the user to type in information.
Group Box	This component is simply used to group other components in logical groupings.
Button	This component is used to let the user click to initiate some action.
Check Box	Check boxes are used when a user can select multiple options.
Radio Button	Radio buttons are used when the user must select one from a list of mutually exclusive options.
Combo Box	The combo box provides a drop-down list the user can select from.
List Box	The list box also displays a list, but it is static rather than drop-down.
Horizontal Scroll Bar	This allows the user to scroll horizontally.
Vertical Scroll Bar	This allows the user to scroll vertically.
Spin	This is used in conjunction with another component, such as an Edit box, to allow the user to increment and decrement the value of the second component.
Progress	This is used to show the user a gradually increasing progress bar.
Slider	With this component, the user can select values by sliding the slider back and forth.
Hot Key	Creates a short cut or hot key.
List Control	Creates a list of items.
Tree Control	This component allows you to display data in a hierarchical tree.
Tab Control	This data allows you to display data grouped logically onto tabs.
Animate	This is used to display short animations.
Rich Edit	This component works much like the Edit box, but has more formatting options.
Date Time Picker	This gives you a drop-down box whereby you can select date and time.
Calendar	The calendar allows you to view a monthly calendar and to select specific dates.
Extended Combo box	This is a combo box that you can also place images in.

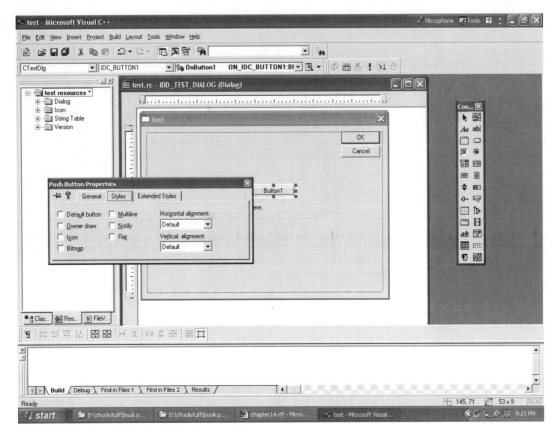

FIGURE 15.17A   The Combo box.

List Box

The List box presents data in a vertical list. It has a number of properties you can set. These properties include vertical scroll, horizontal scroll, and border.

The Dialog Application

The Dialog application has a number of properties you can set. These properties will effect the operation of the dialog itself. The properties tab for the Dialog application is shown in Figure 15.17B.

Several of these properties should look interesting to you. You can set whether your application has a title bar and/or scroll bars, minimize or maximize buttons, and even what type of border it has.

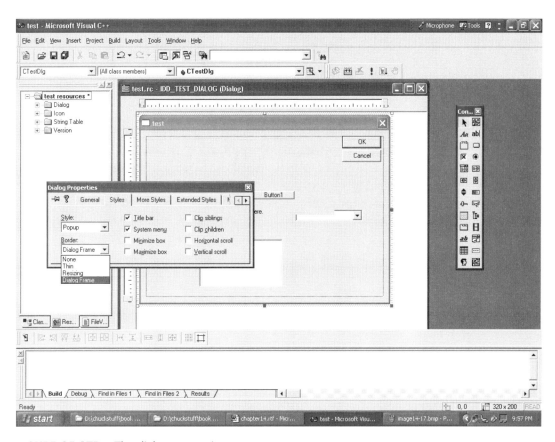

FIGURE 15.17B   The dialog properties.

There are, of course, many other components available to you. We will investigate many of these as this book progresses. You can look any one of them up in the Microsoft *Help* files that come with *Visual C++* if you want more information on them.

BUILT-IN FUNCTIONS

Just as ANSI standard C++ has a number of functions in header files that you can use in your programs, so does *Visual C++*. An exhaustive look at all of these is well beyond the scope of this book. However, we can take a look at some very commonly used functions. To begin with, *Visual C++* supports all the header files, and all their constituent functions that you found in ANSI standard C++. It just adds a few more.

One of the most obvious things that *Visual C++* offers is increased support for strings. In *Visual C++*, you need not deal directly with character arrays. There is a string variable that is actually a class, with many methods in it. The following example will illustrate this to you.

Example 15.2

Step 1: Create a standard dialog application as you did in Example 15.1. You will use the *MFC Application Wizard (exe)* to create a basic dialog applications. The only difference is that you should name it *15_02*.

Step 2: Place four static boxes and four edit boxes on the dialog box as you see in Figure 15.18.

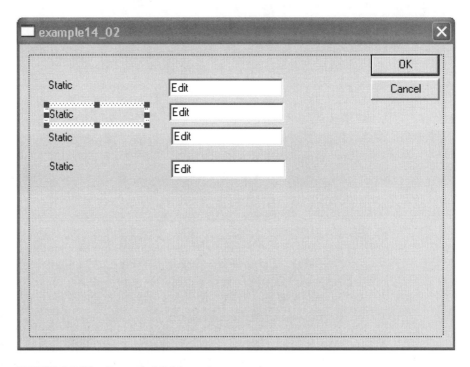

FIGURE 15.18   Example 15.2 layout.

Step 3: Using the right mouse button to view the properties of each of the components, you will change the captions on the static boxes as you see in Figure 15.19. Also add another button with the caption *String Operations.*

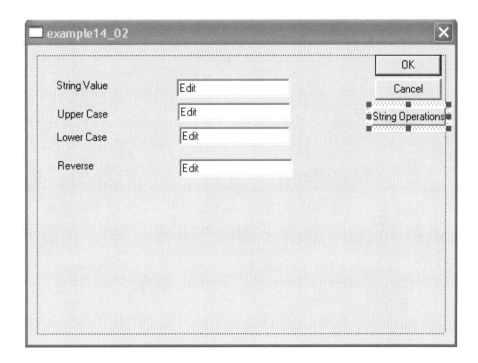

FIGURE 15.19   Labeling the static boxes.

Step 4: With each of the Edit boxes, you will right-click on them and launch the *class wizard.* Using the second tab, member variables, you will assign each of the Edit boxes a string variable as you see in Figure 15.20.

Step 5: Now we just need to add a little bit of code into the string operations button. Simply double-click on the button. (This will add the OnButton1 function to that button.) Then add the code you see here.

```
void CExample15_02Dlg::OnButton1()
{
    // get data from controls to variables.
```

```
                    UpdateData(true);
                    // copy data to the other string variables.
                    m_uppercase =  m_rawstring;
                    m_lowercase = m_rawstring;
                    m_reverse = m_rawstring;
                    // perform string operations.
                    m_uppercase.MakeUpper();
                    m_lowercase.MakeLower();
                    m_reverse.MakeReverse();
                    // get data from variables to controls
                    UpdateData(false);

               }
```

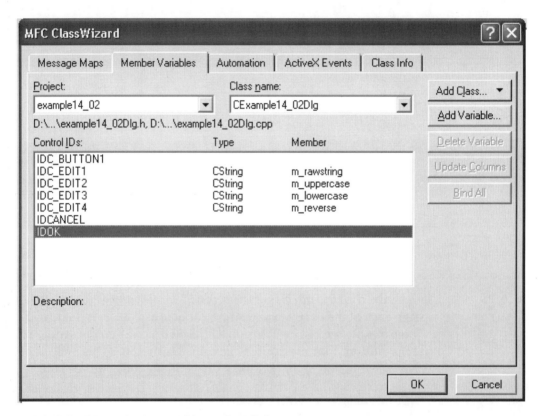

FIGURE 15.20   Assigning variables to the Edit boxes.

Step 6: When you execute the program, you should enter some text into the first `Edit` box then click the `strong operations` button. You will see something much like what is depicted in Figure 15.21.

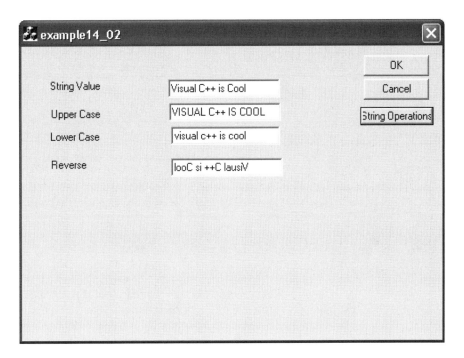

FIGURE 15.21   String operations.

This example should have accomplished several things. First of all, it should have given you another chance to work with the *Application Wizard, Class Wizard,* and with basic *Visual C++* components. Secondly, this example introduced you to some of the functions associated with strings in *Visual C++*. Remember that, in *Visual C++*, the string is not simply an array of characters; rather, it is a class. In fact, it is called the `CString` class. As with any class, it has certain methods. You have just seen the `MakeUpper`, `MakeLower`, and `MakeReverse` methods. Table 15.5 summarizes the main methods of the string.

TABLE 15.5   Methods of the String Class

Method	Purpose
GetLength	This method returns an integer that tells the number of characters in the string.
IsEmpty	This property returns true if the string contains no characters.
Compare	This method compares two strings to see if they are equal.
Mid	This allows you to extract the middle characters from a string.
Left	This method allows you to extract the left side of a string.
Right	This method is used to extract the right side of a string.
MakeUpper	This method returns a string that is the upper case of the string passed to it.
MakeLower	This method returns a string that is the lower case of the string passed to it.
MakeReverse	This method returns a string that is the reverse of the string passed to it.
Find	This method searches to find a series of characters within a string.

HINT!

You should notice that we did not use the *class wizard* to associate a function with the button. We simply double-clicked on the button. Each component has a default function associated with it. If you double-click on the component, it will automatically associate the default component with it.

There are other methods in CString but these are the most commonly used. You will find that *Visual C++* offers a plethora of useful functions that can help you. Covering all such functions, or even a significant portion thereof, is well beyond the scope of this book. However, you will find a few commonly used functions demonstrated in this section of the book, and you can find even more by simply searching through the Microsoft *Help* files that come with *Visual Studio*.

HINT!

Although these *Visual C++* strings are very similar to the standard C++ string and the C style strings, they are not exactly the same.

MESSAGE BEEP FUNCTION

The real focus of *Visual C++* and in *Windows* is to provide a great deal of interaction with the user. You have already seen the message box function used to facilitate this. Your *Windows* operating system also gives you plenty of audio feedback. There are a variety of different sounds for different purposes, including error messages. In *Visual C++*, you have access to many of these system sounds, and you can incorporate them into your own programs. You do this via the message beep function. The message beep function takes a single parameter, an integer constant that tells it which sound to activate. You will then hear the default system sound for that activity. Table 15.6 shows these constants. An example demonstrating the use of the message beep function follows.

TABLE 15.6   Message Beep Constants

Constant	Sound
MB_ICONASTERISK	The system asterisk sound.
MB_ICONHAND	The system hand sound.
MB_ICONQUESTION	The system question sound.
MB_OK	The system default sound.
MB_ICONEXCLAMATION	The system exclamation.

Example 15.3

Step 1: Start the *Application Wizard (exe.)* Use default settings for everything except the name, which should be *example15_03*.

Step 2: Place three buttons on the dialog labeled "Exclamation," "Question," and "OK." It should look like what you see in Figure 15.22.

Step 3: We are now going to place the message beep function call to the appropriate sound in each of those buttons. The code should look like the following.

```
void CExample1_O3Dlg::OnButton1()
{
 MessageBeep(MB_ICONEXCLAMATION);
}
```

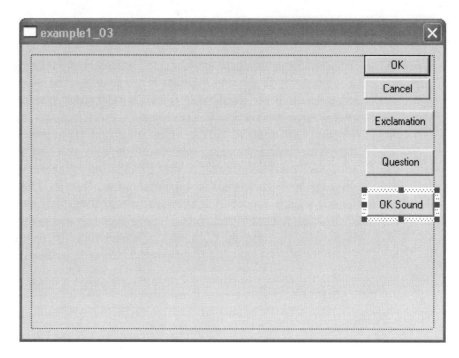

FIGURE 15.22 Message beep demo.

```
void CExample1_03Dlg::OnButton2()
{
MessageBeep(MB_ICONQUESTION);
}
void CExample1_03Dlg::OnButton3()
{
   MessageBeep(MB_OK);
}
```

When you execute the program and click on each of the buttons, you will hear a sound. Of course, if, for some reason, your system has no default sound associated with a given activity, then you will not hear any sound.

MOUSE EVENTS

You have probably noticed that the use of the mouse in *Windows* is very common. Most functions can be accessed via mouse motions and mouse clicking. In fact, *Windows* users are more interested in using the mouse to maneuver through *Windows* than they are in using keyboard commands. It would seem prudent, therefore, for you to gain at least a basic knowledge of how mouse events are handled within *Visual C++*.

As was previously stated, *Windows communicates* by sending out messages. In our case, we are interested in the WM_MOUSEMOVE message. This indicates that the mouse has been moved. There are a number of flags that come with that message to tell you what other keys might have been depressed when the mouse was moved. Table 15.7 summarizes these flags.

TABLE 15.7   Mouse Move Flags

Flag	Purpose
MK_LBUTTON	Indicates that the left mouse button was clicked.
MK_RBUTTON	Indicates that the right mouse button was clicked.
MK_MBUTTON	Indicates that the middle mouse button was clicked.
MK_SHIFT	Indicates that the shift key was clicked
MK_CONTROL	Indicates that the control key was clicked

The other item we will need to examine is the CclientDC. This provides access to what is called the device context. The device context allows us to manipulate the screen to do simple drawings. It is a class that you create an instance of in order to give you access to the device context. Its constructor takes a single parameter. That parameter is the keyword this. Recall that this refers to the current instance of a class you are in. Let's look at an example that utilizes the CclientDC and mouse functions to do a simple drawing. This should help you become acquainted with mouse events.

Example 15.4

Step 1: Start *Visual C++* and run the *app wizard* to create a basic dialog application with the default values, except for the name, which should be *example15_04*.

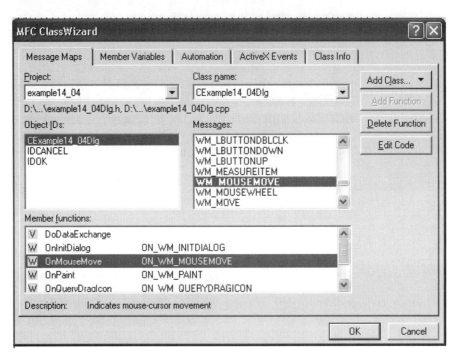

FIGURE 15.23   Adding functions for MOUSEMOVE.

Step 2: Open the *class wizard* (Figure 15.23). On the first tab in the mes-sages box, choose WM_MOUSEMOVE. Now click add function, then click OK. Then click the Edit Code button.

Step 3: Add the following code.

```
    if((nFlags & MK_LBUTTON) == MK_LBUTTON)
{
 CClientDC dc(this);//device context with this dialog
//class as a reference.
      dc.SetPixel(point.x,point.y,RGB(0,0,0));
 }// end of if
```

At this point you have a working, if somewhat uninteresting, piece of code. You should notice the `if` statement. It is using a single ampersand (&), which is a bitwise AND operation. You see the `nFlags` is a variable that holds an integer representing whatever button is currently being clicked. The bitwise AND operation compares this to the `MK_LBUTTON` integer value to see if that is the button that is being clicked. You can run the code now. A dotted line will follow as you move the mouse around the screen, as you see in Figure 15.24.

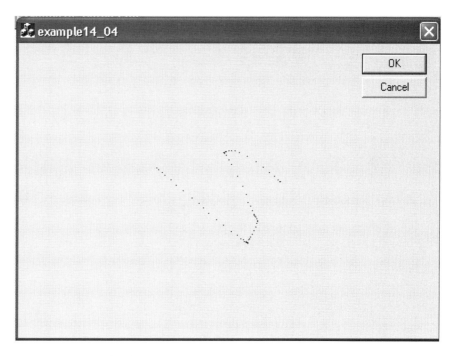

FIGURE 15.24   The first implementation of MOUSEMOVE.

However, now we would like to beef this up a bit.

Step 4: Use the class view and right-click on the Dialog's class and then add a member variable and make it private and an `int`. Add another variable, also an `int` and private. This should look similar to Figure 15.25.

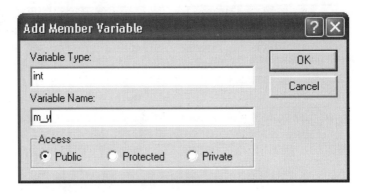

FIGURE 15.25 Adding member variables.

Step 5: Replace the code you had originally in the onMouseMove function with the following code.

```
// Check to see if the left mouse button is down
if ((nFlags & MK_LBUTTON) == MK_LBUTTON)
{
// Get the Device Context
CClientDC dc(this);
// Draw the pixel
// dc.SetPixel(point.x, point.y, RGB(0, 0, 0));
// Create a new pen
CPen lpen(PS_SOLID, 16, RGB(255, 0, 0));
// Use the new pen
dc.SelectObject(&lpen);
// Draw a line from the previous point to the
// current point
dc.MoveTo(m_X, m_Y);
dc.LineTo(point.x, point.y);
// Save the current point as the previous point
m_X= point.x;
m_Y = point.y;
}
CDialog::OnMouseMove(nFlags, point);
```

When you run the application this time, you will be able to draw wide, solid lines, like those shown in Figure 15.26.

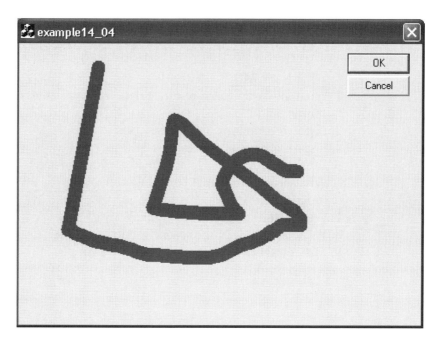

FIGURE 15.26   More MOUSEMOVE functions.

SUMMARY

This chapter has introduced you to *Windows* programming with *Visual C++*, an exciting but daunting area of programming. Many programmers feel overwhelmed when they first encounter *Windows* programming with *Visual C++*. There are a number of complexities to come to grips with. The important items for you to grasp right now are actually quite few. To begin with, you should basically understand what the Microsoft *Foundation Classes* are. If you understand what they are and basically how they are used, then you will be able to use reference material such as books and Microsoft *Help* files to find the specific class you need to perform a specific operation.

You should also become comfortable with the *MFC Application Wizard* and with the *Class Wizard*. If you are comfortable with these two wizards, your future *Visual C++* programming endeavors will be greatly simplified. It would be wise to do a few more simple applications using these wizards just to make certain that you are totally comfortable with them.

REVIEW QUESTIONS

1. What is the *MFC*?
2. What method is used to move data from variables to the associated components?
3. List five common components used in *Visual C++*.
4. List 4 methods of CString.
5. What two functions do all *Windows* programs have?
6. What header file must be included in order to use the message box function?
7. What is the purpose of the *Class Wizard*?
8. What is the title of the second tab in the *Class Wizard*?
9. List at least two constants for the message beep function.

More Windows Applications with Visual C++

IN THIS CHAPTER
•••••••••••••••

- Additional Components
- Menus
- Icons
- Modifying the about Dialog
- SDI and MDI Applications

Chapter 15 introduced you to *Windows* programming with *Visual C++*. You should have learned the basics of *Windows* programming, specifically with dialog applications. This chapter will build on that basis, and start showing you how to create more advanced applications. This will include using more components, creating menus, and modifying the about dialog. This chapter builds on the foundation of the last chapter so it is imperative that you understand the material from chapter 15. If you do not, then please go back and review that chapter.

This chapter will also give you a brief introduction to SDI and MDI applications. SDI applications are Single Document Interface applications.

Essentially that is an application that can view one document at a time. If you have ever used *Notepad* or *Wordpad* in Microsoft *Windows,* then you have used Single Document Interface applications. MDI applications are Multiple Document Interface applications. If you have used Microsoft *Word* or Microsoft *Excel*™, then you have used Multiple Document Interface applications.

ADDITIONAL COMPONENTS

The first thing we wish to concentrate on in this chapter is how to use additional components. This means you will get more experience using the *application wizard,* dialog applications, and the *class wizard.* You have already seen the use of Edit boxes and buttons. You should be at least moderately comfortable with these components. Another important component to become familiar with is the radio button. A radio button is used to allow a user to make a single choice between several mutually exclusive choices. For example if you wanted a user to tell you if his or her age was between 18 and 30, 31 and 40, 41 and 64, or over 65. These are mutually exclusive choices. Only one choice can be true. For situations like this, a radio button is a good choice. In this next example, you will see how to use the radio button, and you will also get another chance to work with the Button and Edit boxes. Table 16.1 lists some option button methods.

TABLE 16.1   Option Button Methods and Properties

Method/Property	Purpose
OnClick	This method is executed when the radio button is clicked.
Style-Flat	This property makes the button appear flat rather than three-dimensional.
Style–Icon	This property is used to associate an icon with a button.
Style-Bitmap	This property is used to associate a specific bitmap with a radio button.
Caption	This property sets the caption to display. This is not the same as the ID. The ID is what your code will use to refer to the component. The caption is what your users will see.

Example 16.1

Step 1: Start a new dialog application using the *application wizard* and default values. This will follow the same process you followed in the previous chapter.

Step 2: Place four `radio buttons`, two `Edit boxes`, and one `button` on the dialog. It should look like what you see in Figure 16.1.

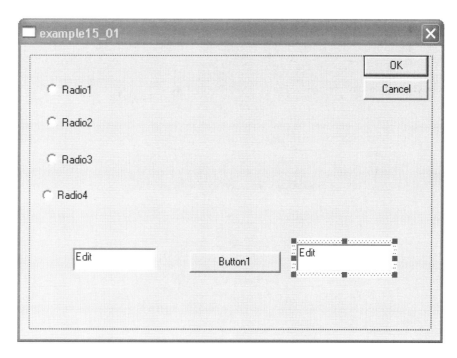

FIGURE 16.1   Dialog layout for Example 16.1.

Step 3: On each of the components, you will right-click the mouse to get to its properties. You will then change their captions to match the captions depicted in Figure 16.2.

Step 4: Now you will need to use the *class wizard* to associate a variable with each of the `Edit boxes`. You will do this by right-clicking on a component and choosing *class wizard*. You then select the `member variables` tab and add a variable. This is shown for the first `Edit box` in Figure 16.3.

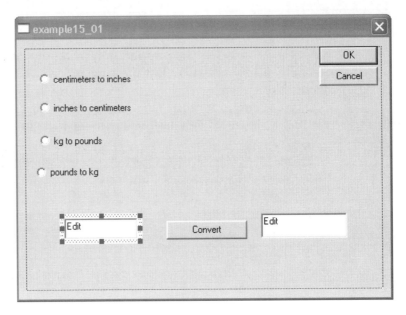

FIGURE 16.2   Component captions.

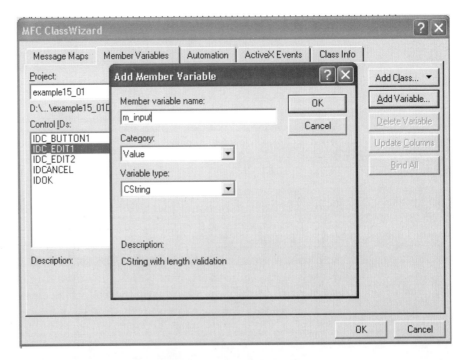

FIGURE 16.3   Providing variables for the components.

HINT!

The Edit boxes should be given a float type variable as shown in Figure 16.4.

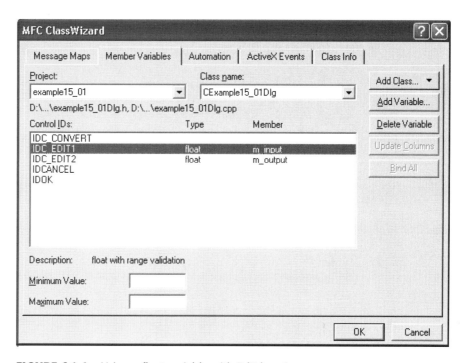

FIGURE 16.4   Using a float variable with Edit box 1.

You will also use the properties dialog to change the names of the button and the four radio buttons. The button's ID (the ID is another word for its name) will be changed to IDC_CONVERT , this is shown in Figure 16.4. The option buttons' names will be changed to match their captions. The following names will be used.

```
IDC_CMTOINCH
IDC_INCHTOCM
IDC_KGTOLB
IDC_LBTOKG
```

As was mentioned in Chapter 1, proper naming of variables can make your code much more readable and, hence, maintainable. Now you can double-click on any component on the dialog and scroll to the top. We are going to create a global variable and some global constants.

```
int OptionSelected;
const int CMTOINCH = 1;
const int INCHTOCM = 2;
const int LBTOKG = 3;
const int KGTOLB = 4;
```

Step 5: Now you are going to double-click on each of the radio buttons, thus creating a function that matches that radio button. You will then add a single line of code that will set the OptionSelected variable to a constant that matches the radio button clicked. This should look like what you see here.

```
void CExample16_01Dlg::OnCmtoinch()
{
    OptionSelected = CMTOINCH;
  }
```

Step 6: Now you can double-click on the convert button and add the following code.

```
void CExample16_01Dlg::OnConvert()
{
    float answer;
 UpdateData(TRUE);
 switch(OptionSelected)
 {
 case CMTOINCH:
   answer = m_input * .3937007f;
      break;
 case INCHTOCM:
   answer = m_input * 2.54f;
      break;
 case LBTOKG:
   answer = m_input *.454545f;
      break;
 case KGTOLB:
   answer = m_input * 2.2f;
      break;
 default:
```

```
    AfxMessageBox("Please make a selection");
    }//end switch
m_output = answer;
UpdateData(FALSE);
}
```

Step 7: Now compile and execute your code. You should see something like what is shown in Figure 16.5.

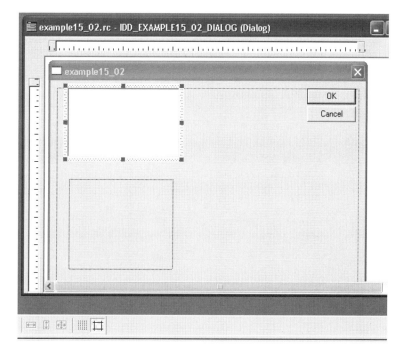

FIGURE 16.5   Unit conversions.

Let's take a look at exactly what is happening in this application. Essentially we use the onClick function of the radio buttons to set a variable named OptionSelected. That variable's value will tell us which conversion was selected. We then simply retrieve the value from the first edit box. Next there is a simple switch case statement that uses the value in OptionSelected to choose the appropriate formula to execute. Finally, we place the answer into the second Edit box. This is a relatively straightforward piece of code. The only thing new in this program is the use of the radio buttons.

Two other important components you need to be comfortable with are the list box and combo box. A list box is used to present a list, which you select from. All items are presented at a single time. A combo box is much like a list box, except only a single item is initially displayed and then the other items show as a drop-down list. These two components are commonly used in *Windows* programming, so it would behoove us to examine them in some detail. The next example should help you to become better acquainted with these two components. First, however, you should study Tables 16.2 and 16.3 to see the methods and properties of the list box and the combo box.

TABLE 16.2   List Box Methods and Properties

Methods/Properties	Purpose
ON_LBN_DOUBLECLICK	This method is executed when an item in the list box is double-clicked.
ON_LBN_SETFOCUS	This method is executed when the list box receives focus.
Sort	This property causes the list box to automatically be sorted.
Border	This property causes the list box to have a border.
Selection	This property determines what kind of selection the user can make. The options are: Single, Multiple, Extended, and None.

TABLE 16.3   Combo Box Methods and Properties

Methods/Properties	Purpose
Sort	This property causes the items to be sorted.
Vertical Scroll	This property causes a vertical scroll bar to be placed in the combo box.
Type	This property allows you to select what type of combo box you wish. The choices are simple, drop-down list, and drop-down combo.
UpperCase	This property causes all items in the combo box to be displayed in all uppercase.

Example 16.2

Step 1: Start a new dialog application, just as you did in the previous example.

Step 2: Place a list box and a combo box on the dialog.

HINT!

 Remember that the area you draw the combo box to fill is the area it will drop down to when you click on it. That's why you should draw it out to a significant size, as you see in Figure 16.6.

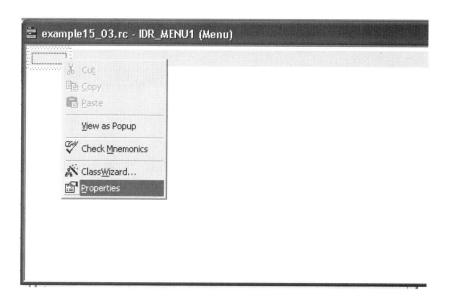

FIGURE 16.6   The combo box.

Step 3: You will also place Edit box on the screen. It should look like what you see in Figure 16.7.

Step 4: Next, you will need to give a variable to each of the components. Each variable will be a CString variable type, as you see in Figure 16.8.

Step 5: Next, we add a few items to the combo box. You do this by right-clicking on the combo box and selecting properties. You then choose the data tab. After you enter each item, press the CTRL – ENTER keys. We are

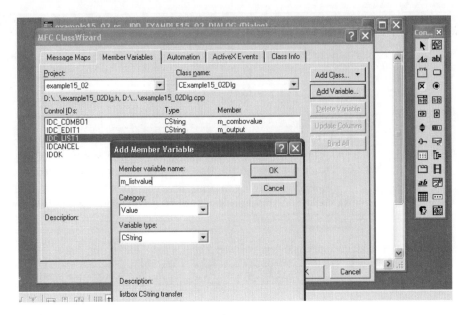

FIGURE 16.7 The dialog component layout.

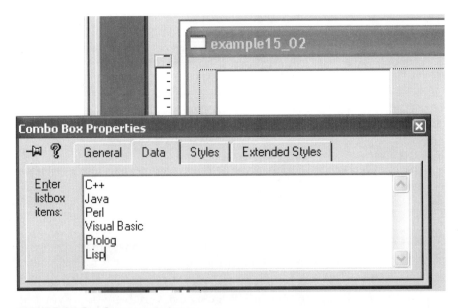

FIGURE 16.8 Component names.

going to enter the names of several popular programming languages, as you see in Figure 16.9.

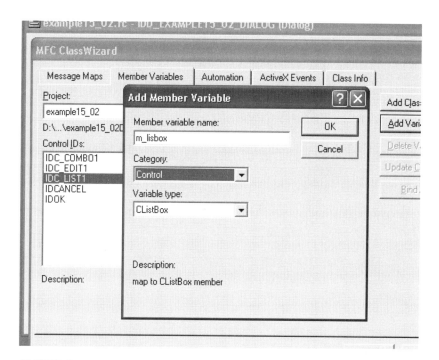

FIGURE 16.9   Combo box data.

When you are done, simply press the enter key to close the property window.

Step 6: Add one more variable. This one will be for the list box, but we will change its type from value to control. This means it will represent the actual component. This should look similar to Figure 16.10.

Now we will need to add code to the OnInitDialog function to add items to the list box. You get to that function via the *class wizard*, as you see in Figure 16.11.

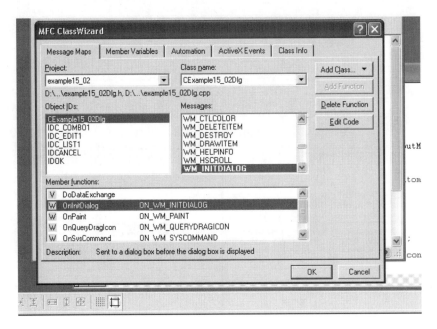

FIGURE 16.10 A variable for the list box.

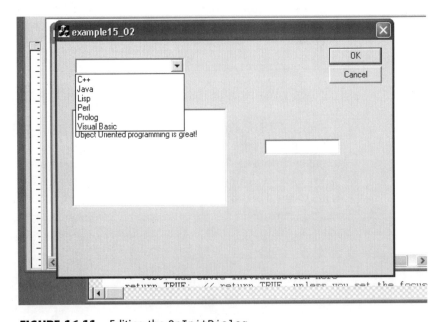

FIGURE 16.11 Editing the OnInitDialog.

Step 7: Now, after you have clicked on edit code and are ready to edit the OnInitDialog function, you will simply add a few lines of code.

```
m_listbox.AddString("C++ is awesome");
m_listbox.AddString("Bjarne Stroustrap is cool");
m_listbox.AddString("Object Oriented programming is
great!");
```

Now, if you will compile and run this program, you should see something much like what is depicted in Figure 16.12.

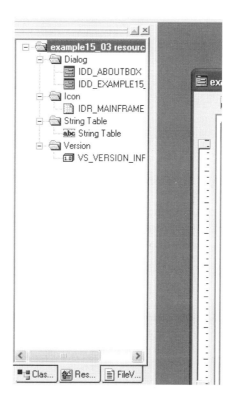

FIGURE 16.12 Running the dialog application.

This application, while admittedly simple, illustrates how to use the combo box and the list box. There are, however, a few items that need some further explanation. The OnInitDialog is a function that is executed as soon as the dialog is loaded into memory. It's a good place to do any code that you want executed before the user can do anything else. You can think of this function as being similar to a class's constructor. For those of you with a background in *Visual Basic*, the OnInitDialog is analogous to the form load event. You should also notice that there are different types of variables that can be associated with a given component. The value variables are standard data types that represent the data the component holds. The control variables represent the component itself, and give you access to that component's methods.

These are just two more components you have available to you in *Visual C++*. Covering every component available to you is beyond the scope of this book. However, the Edit box, button, list box, combo box, and static text are perhaps the most commonly used. The *Help* files that ship with *Visual C++* give you a nice overview of how to use other components in *Visual C++*.

MENUS

Most *Windows* applications also support drop-down menus that allow you to select particular actions you wish to perform. As you might guess, *Visual C++* facilitates this as well. Before we can start creating menus, you will need a brief introduction to resources. *Visual C++* allows you to incorporate various resources into your program. These resources are often of a graphical nature. You may not have noticed that the tab on the lefthand side has three small tabs on it. The first is class view and the last is the file view. The one that is of most concern to us at this point is the middle one, the resource tab. Any graphical elements of your program, including the dialog itself, are resources. If you look at the resources in any standard dialog application, you will see something much like what is depicted in Figure 16.13.

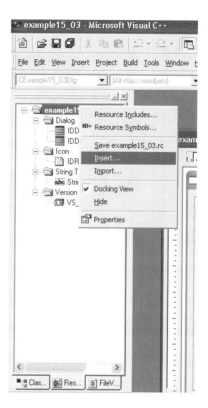

FIGURE 16.13   Resource tab.

There are a number of resources listed in this window, including the dialog itself. Some of these we will return to later in this book. For now, our focus will be on how to add menus to the resource tab. Fortunately, this is a relatively simple task. If you will right-click on the top of the resource list, you will be presented with an opportunity to insert new resources, including menus. This is shown in Figure 16.14.

As you can see, there are several types of resources that one can insert from this screen. Let us take a look at what happens after you select menu. When you choose to insert a new menu, you will see a screen much like the one depicted in Figure 16.15.

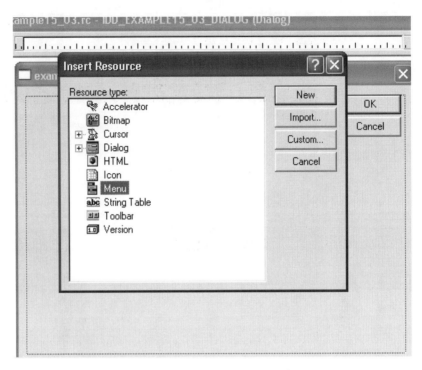

FIGURE 16.14   New menus.

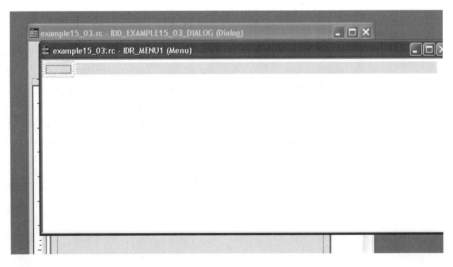

FIGURE 16.15   A new menu item.

The next step is to right-click on the menu and choose properties, as is shown in Figure 16.16.

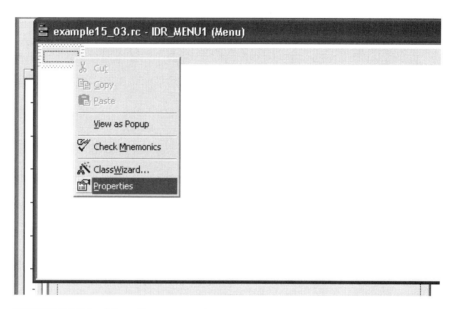

FIGURE 16.16   Menu item properties.

You will then be able to assign a caption to the menu, decide whether it is a checked menu, if it is grayed out, and so on, as shown in Figure 16.17.

Now that you know how to create a menu, there is still one more issue to address. That issue is, how do you connect that menu to an application? That connection does not occur automatically. If you run your application now, it will not have a menu. Fortunately, connecting the menu to the dialog is quite simple. If you will right-click on the dialog itself, and select properties, you will see something like what is depicted in Figure 16.18.

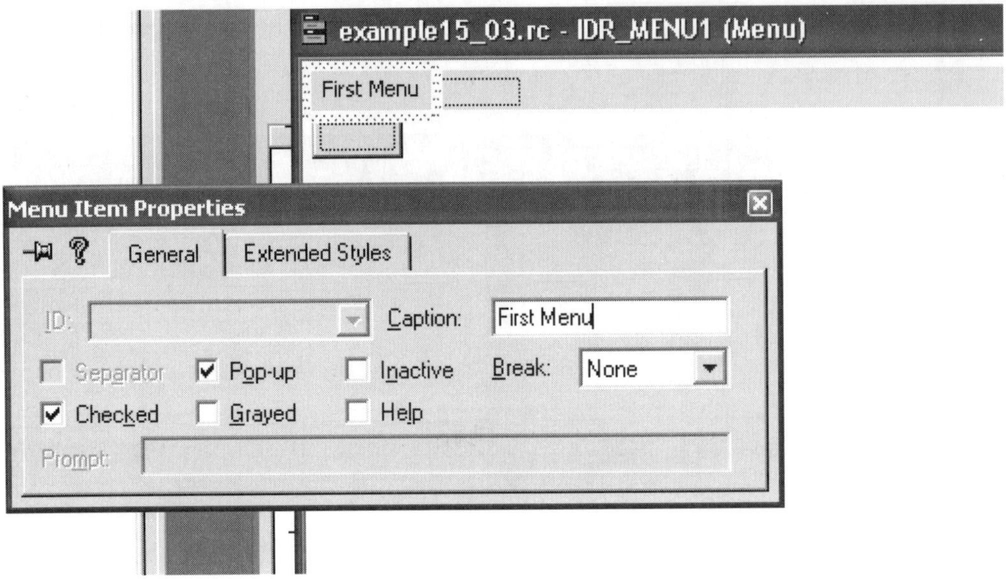

FIGURE 16.17   Selecting menu item properties.

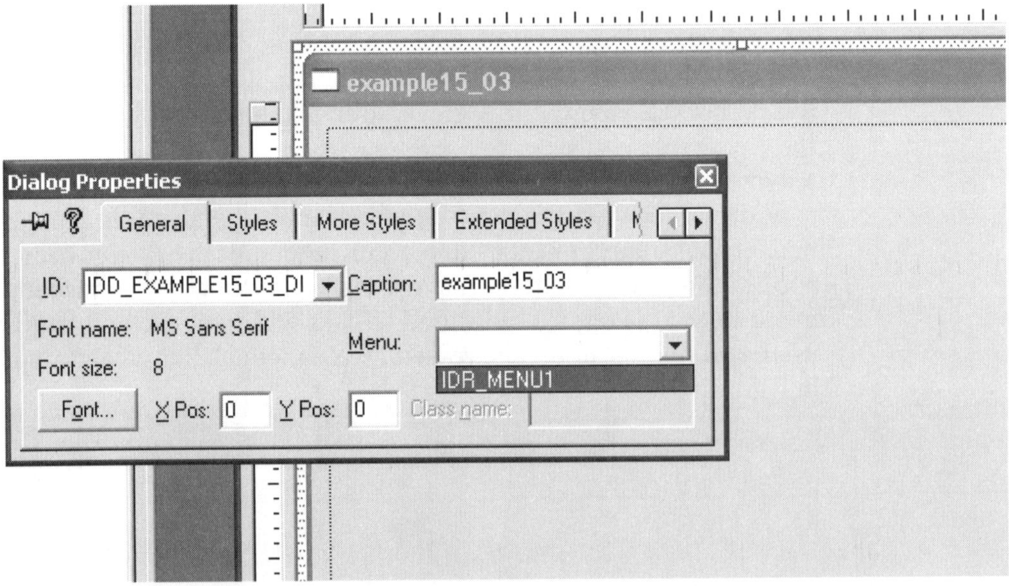

FIGURE 16.18   Associating a menu with a dialog.

HINT!

Make sure that you are right-clicking on the dialog and not on an individual component on the dialog.

You can see that you can now associate any given menu with any dialog. Creating menus and associating them with dialogs is a relatively straightforward operation that you can accomplish in just a few steps. You will find that adding drop-down menus to your applications makes them much more user-friendly. Let's look at an example that shows the use of the drop-down menu, coupled with math functions that we saw earlier in this book.

Example 16.3

Step 1: Create a standard dialog, just as you have been doing with the previous examples.

Step 2: Add two Edit boxes and two static boxes, as you see in Figure 16.19.

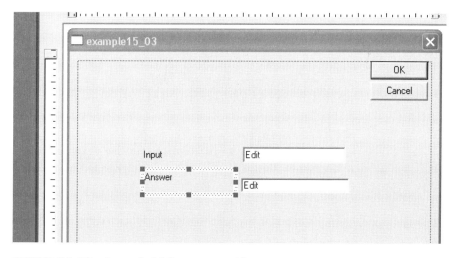

FIGURE 16.19   Example 16.3 component layout.

Step 3: Now use the *class wizard* to associate a `float` variable with each of the `Edit` boxes, as is shown in Figure 16.20.

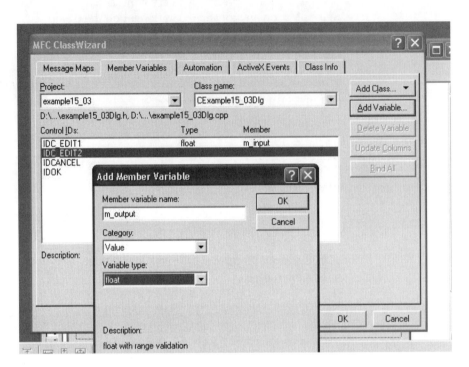

FIGURE 16.20   Naming the Edit boxes.

Step 4: We will now add a series of menus. You should recall from the previous discussion that you will need to select the `resource tab`, and right-click on it to add a menu. If you continue to right-click on the menu, just below the previous item, you can add several items in one menu, as shown in Figure 16.21.

We will add four menu items: `Math Operations`, `SQRT`, `Cube`, and `Square`, as shown in Figure 16.22.

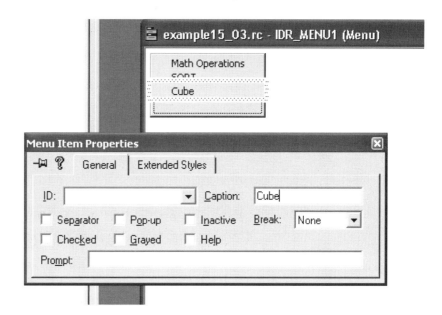

FIGURE 16.21   Adding multiple menu items.

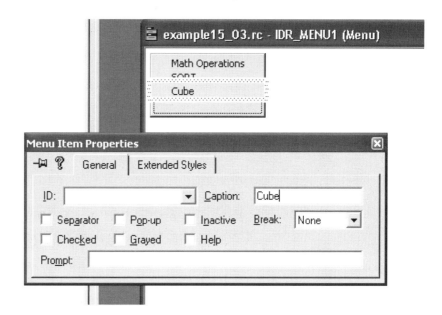

FIGURE16.22   Menu items for Example 16.3.

Step 5: Next, we will associate that menu with the dialog. Remember that you accomplish this by right-clicking on the dialog itself and choosing `properties`.

Step 6: It is now time to use the *class wizard* to add some functionality to our menu items. When you right-click on the menu and select *class wizard*, the wizard will prompt you for some information. Simply choose the same settings as you see in Figures 16.23 and 16.24.

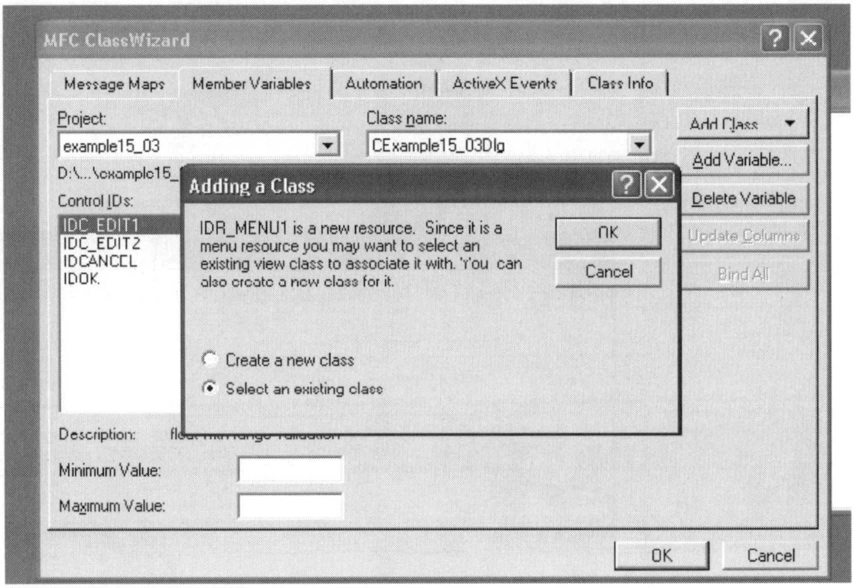

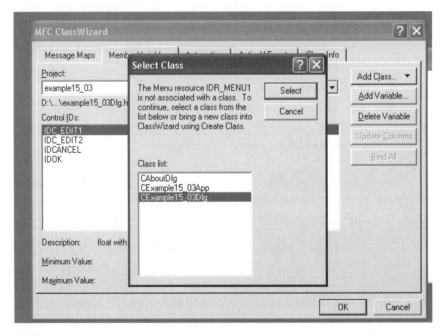

FIGURE 16.23   Launching the class wizard.

FIGURE 16.24   Class wizard continued.

Step 7: Now we can add functions to our menu items. To do this, you use tab one of the *class wizard* (Message Maps), select the menu item, then select the word command in the right pane. This is shown in Figure 16.25. You can then choose add function, and add an OnMenuItem function as shown in Figure 16.26.

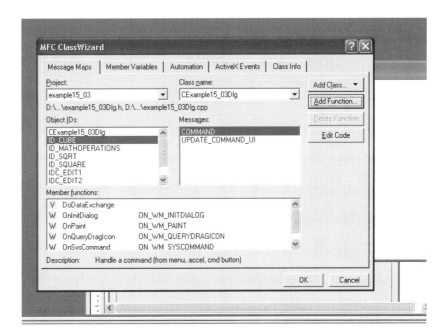

FIGURE 16.25   Selecting a menu item.

You will need to add a function for each of these menu items. Thus, you will have an OnSquare, OnCube, OnMathOperations, and onSQRT function. If you then choose to edit any of these functions (and you should!), a window will pop-up that displays the code for all these functions. Your first order of business is to scroll to the top of that source file and add the include statement for the cmath header file.

```
#include <cmath>
```

Step 8: Scrolling back down to the functions, we are going to add just a few lines of code as you see here.

```
void CExample16_03Dlg::OnCube()
{
```

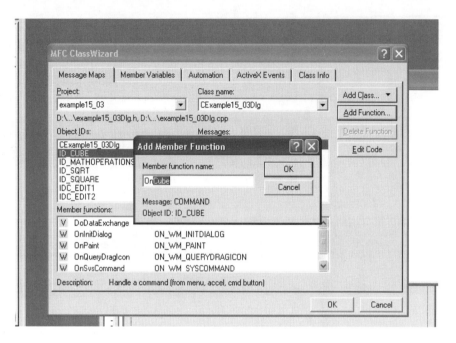

FIGURE 16.26   Adding a function.

```
    UpdateData(TRUE);
    m_output = pow(m_input,3);
    UpdateData(FALSE);

}
void CExample16_03Dlg::OnSqrt()
{
    UpdateData(TRUE);
    m_output = sqrt(m_input);
    UpdateData(FALSE);
}
void CExample16_03Dlg::OnMathoperations()
{
    AfxMessageBox("Please select one of the math
 operations");
}
void CExample16_03Dlg::OnSquare()
{
     UpdateData(TRUE);
     m_output = pow(m_input,2);
     UpdateData(FALSE);
     }
```

Step 9: Now it's time to execute this program. You should see something much like what is depicted in Figure 16.27.

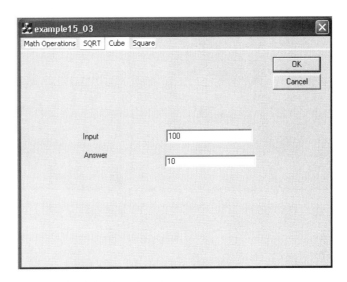

FIGURE 16.27   Running menu programs.

When you enter a number into the first Edit box, and select one of the menu options, a mathematical operation is performed on that number, and the answer is placed in the second Edit box. This application, while simple, should illustrate to you how to utilize the menus, and how to incorporate standard C++, such as the functions in cmath, with *Visual C++*.

ICONS

Most *Windows* applications are represented by an icon. That icon image will show up when you see the program in *Windows Explorer*, put a shortcut on the desktop, or on the upper lefthand corner of the applications main screen. If you had not noticed, the *Windows* applications you have created in this chapter and the last, have icons. However, they are rather unattractive ones. Fortunately, *Visual C++* makes it very easy for you to edit existing icons, or to add new ones.

If you will take a look at the last example, you will find a section in resources, where a default icon was created. It has the letters MFC on it, see Figure 16.28.

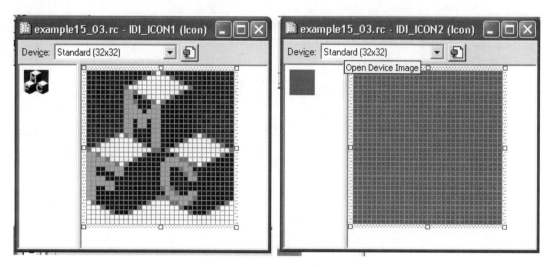

FIGURE 16.28   The default icon.   **FIGURE16.29**   Drawing your own icon.

Visual C++ allows you to edit that image. As you can see, there are tools available much like you would find in any basic graphics software package, such as Microsoft *Paint*. You can use these tools to edit the existing image or create an entirely new image. To create an entirely new image, you right-click on the icon heading under resources, and then choose insert icon. You will get a blank icon, like the one shown in Figure 16.29, and you can then draw anything you like.

However, if your artistic skills are somewhat lacking, you may wish to use some preexisting icon. This is actually quite easy to do. If you right-click on the icon heading in resources and choose import, you will be presented with a dialog box, much like the one shown in Figure 16.30, and you can import any icon that happens to be on your computer.

Whether you import an existing icon, edit the default icon, or create your own icon, the problem still remains of how to associate that icon with a particular dialog. You will notice that the default icon provided for you by the application wizard, has the ID of IDR_MAINFRAME. If you will simply delete that icon and rename yours as IDR_MAINFRAME, your icon will be used. Changing your icon's name is shown in Figure 16.31.

Some programmers ignore details such as the application icon. You should, however, pay close attention to the details of your graphical user interface. Every detail creates an impression on the people using your application, and you want them to see a professional and polished product.

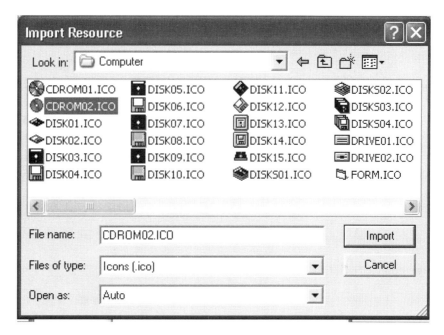

FIGURE 16.30   Importing existing icons.

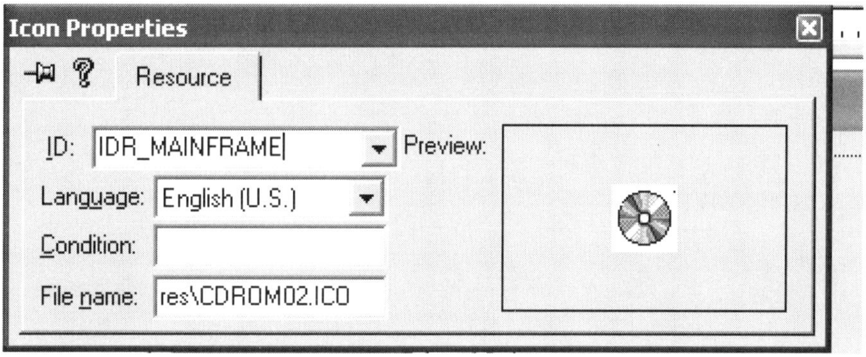

FIGURE 16.31   Changing your icon's ID.

The real advantage to resources in *Visual C++* lies in how they are handled when you compile your application. All the resources are actually compiled into the executable. This means that when you distribute your program, you do not have to worry about distributing the icons as well as the .exe file. The icons are already included in the executable.

CUSTOMIZING THE ABOUT DIALOG

You may have noticed that each time we use the *application wizard,* an about dialog box is also generated for you. If you had not noticed this, then you should probably take a second look at some of the previous examples. In either case, an about dialog is just a simple window that gives the user some information about the application. For example, if you open Microsoft *Internet Explorer* and select *Help* and *About*, you will see an image much like the one shown in Figure 16.32.

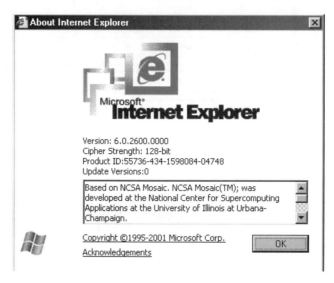

FIGURE 16.32   Microsoft Internet Explorer help screen.

The *Visual C++ application wizard* also creates an about dialog for you. Unfortunately, it creates a rather plain-looking and generic about dialog. The standard help dialog generated for Example 16.3 is shown in Figure 16.33.

This is not something that will be particularly impressive to users of your program. And contrary to the opinions of many programmers, the professional appearance of your application will strongly influence the impression people have of the quality of your program. It would probably be helpful to dress this program up a bit. Fortunately, this is quite simple to do, using standard techniques that you already know.

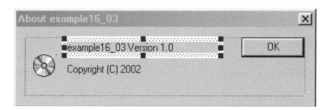

FIGURE 16.33   The generic about dialog box.

The first thing you will want to do is to locate the about dialog. If you look on the resource tab you will see both the main dialog and the about dialog. You can then double-click on the about dialog and it will be opened up in the main window.

Now that you have opened up this about dialog (Figure 16.34), you can easily alter the various properties of it to give it a more polished appearance. The first thing we would like to do is to change the caption of the dialog. To do that, simply right click anywhere on the dialog (not on one of the controls on the dialog), and choose properties (Figure 16.35).

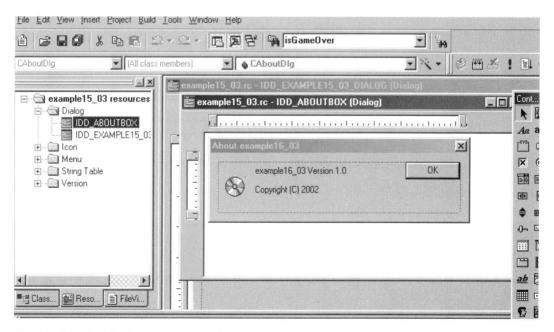

FIGURE 16.34   Viewing the about dialog.

FIGURE 16.35 The properties window.

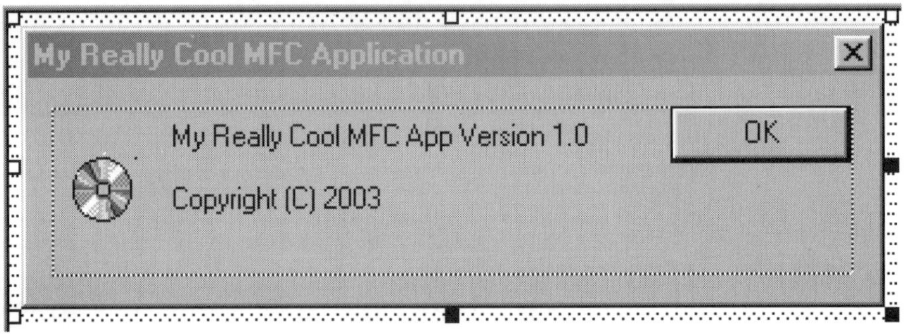

FIGURE 16.36 The captions on the about dialog.

The first thing you will need to do is to change the caption. Simply write the following: My Really Cool MFC App (Figure 16.36).

Now you will change the captions of the static boxes on the about dialog. The generic captions provided by the *application wizard* are somewhat lacking in visual appeal.

Last, but by no means least, we want to change the icon displayed here. We want to do this for two reasons. First of all, we want to change the icon because the *application wizard*, by default, gives the about dialog the basic icon that is assigned to the main dialog. We want to use something a bit more creative. Secondly, we want to change this icon so that you might have more practice working with icons.

Our first step will be to add a new icon. We are going to actually create our own icon for the about screen. Although many readers may not be artistically inclined, the drawing will be easy to accomplish. (This is important, as the author of this book is not particularly artistically talented!) If you will recall, you add a new icon by right-clicking on `icons` in the `resource tab` and choosing `insert new icon` (Figure 16.37).

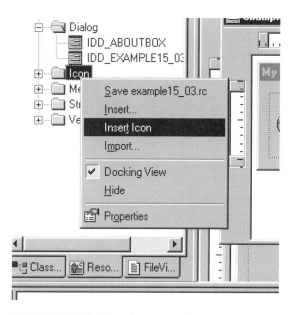

FIGURE 16.37   Inserting a new icon.

You now have a blank format on which to draw any image you wish. Now many readers may be hesitant to attempt to draw their own icons; however, there are a number of tools that will help you. We are first going to utilize the circle tool to draw a plane circle on the screen with a blue color. This process is shown in Figures 16.38 and 16.39.

Now you will have a circle image on the screen to manipulate. Next, we will simply change the color to red and then draw a smiley face on the image (Figures 16.40 and 16.41). You will need to switch back to the pencil tool first.

The drawing is relatively simple, because you can color in a single pixel at a time. You do this by simply clicking in a square with your mouse. Clearly our image is not going to hang in any art galleries. However, it will be a bit more creative than the default icon. Now our

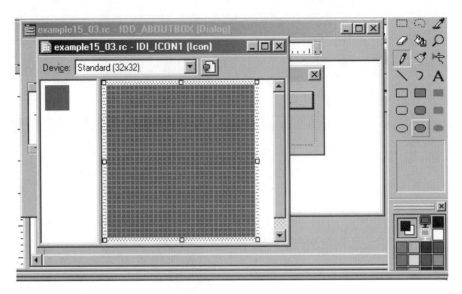

FIGURE 16.38   Picking the circle tool.

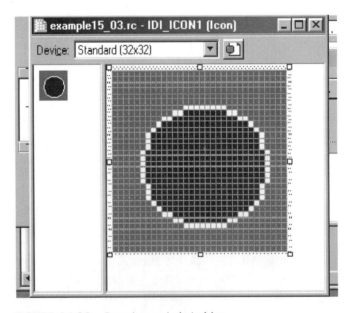

FIGURE 16.39   Drawing a circle in blue.

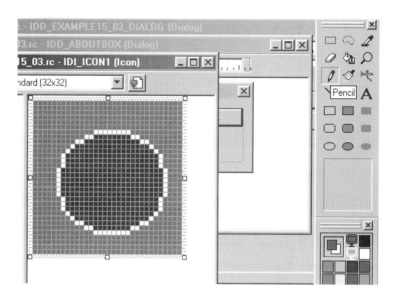

FIGURE 16.40   The pencil tool.

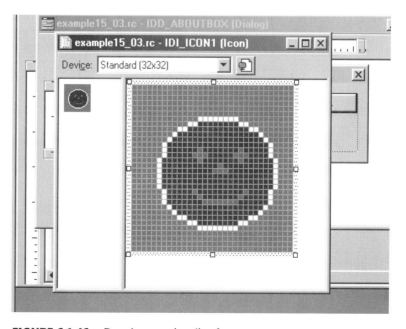

FIGURE 16.41   Drawing a red smiley face.

only issue is to associate the icon with the image. When you added the icon, it was given a name ID_ICON1. Now we have to associate that name with the about dialog. If you will right-click on the icon that is currently on the about dialog, you can select any icon in your project to associate with the about dialog (Figure 16.42).

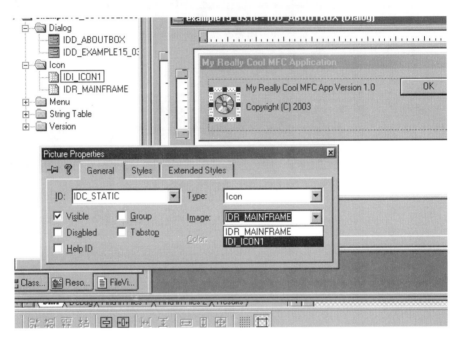

FIGURE 16.42   Changing the about dialog icon.

Now when you execute the program and view the about screen, you will see the icon that you just created! This little piece of artwork shows how you might go about altering icons for use in your applications. Custom icons are often better than pre-made icons.

SDI AND MDI APPLICATIONS

As mentioned at the beginning of this chapter, SDI applications are Single Document Interface applications, and MDI applications are Multiple Document Interface applications. The real difference is simply how many documents you can have simultaneously open in your application.

Also recall that Microsoft *Word* is a good example of a MDI application, where as MS *Notepad* is an example of a SDI application. The one issue that was not previously addressed is defining exactly what a document is. You might assume that it is something like a text document or Microsoft *Word* document, and that assumption would be correct. However, a document can be other things as well. A document can be a paint screen, like in Adobe *Photoshop*™. *Photoshop* is a MDI application; you can view several images at one time. Each image is, in effect, a document.

The *application wizard* (Figure 16.43) allows you to easily create the basic framework for either an SDI application or a MDI application. The entire framework, including the menus, is created for you. You then simply have to add in the specific application code where needed. Because both SDI and MDI applications are rather open-ended application types with a plethora of possible applications, this chapter will simply walk you through the wizard. You can then use various resources in the appendices at the end of the book to find specific ideas you might want to implement in a MDI or SDI application.

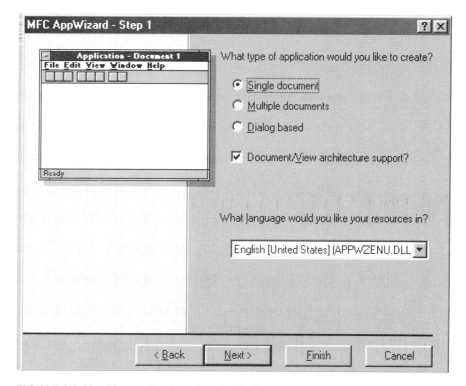

FIGURE 16.43   The *application wizard* with SDI apps.

The SDI framework

Step 1: Start the *application wizard*, you will choose the SDI application type (Figure 16.44).

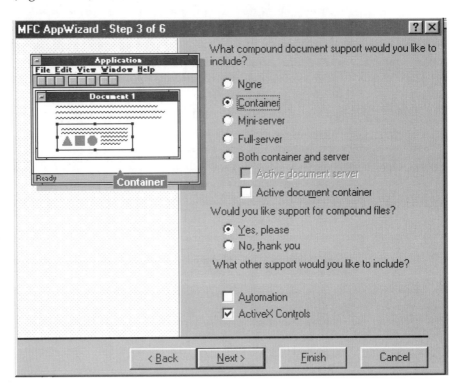

FIGURE 16.44   Choosing the container option.

For the next few steps of the *application wizard*, you will simply keep default settings—until you get to Step 3 of 6 on the *application wizard*. Here you will make one single change. You will select the container option.

What this does is allow your application to contain documents, thus the term *container*. Clearly, your application will have to contain other items to show text documents, image documents, or any other type of document you might be interested in.

Throughout the rest of the *application wizard*, you will select the default options, until you reach the final screen of the *application wizard*. However, it would be useful for you to take some time to examine Step 4 of 6 (Figure 16.45A).

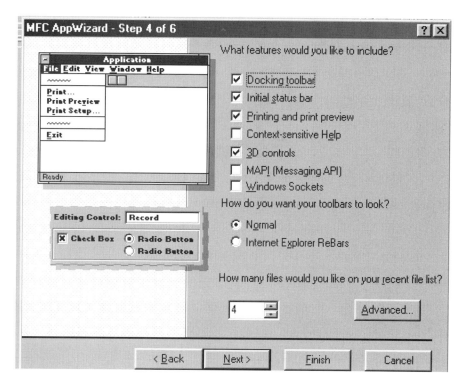

FIGURE 16.45A   Step 4 of 6 on the *application wizard*.

You have a number of useful items you can choose from here. Any of the items selected will be automatically created for you by the *application wizard*. Each of these items is quite useful for you. For example, recall the work done to add in a menu? Well, at this stage in the *application wizard*, if you select the menu box, then a menu will be created for you, with all the standard options you are used to seeing in menus (file, view, edit, etc.). You can also choose to have printing/print preview support, as well as a docking tool bar created for you. This is all very useful and takes care of the work for you.

When you get to the final screen of the *application wizard,* you will notice that there is a section titled *Base Class*. Recall from the previous discussions of object-oriented programming that a base class is the class that is inherited from. Inheritance is a major part of the Microsoft *Foundation Classes*, and, hence, *Visual C++*. Every application has some base class. If we simply change our base class to the CrichEditView, as shown in Figure 16.45B, your application will support actual writing and editing of text.

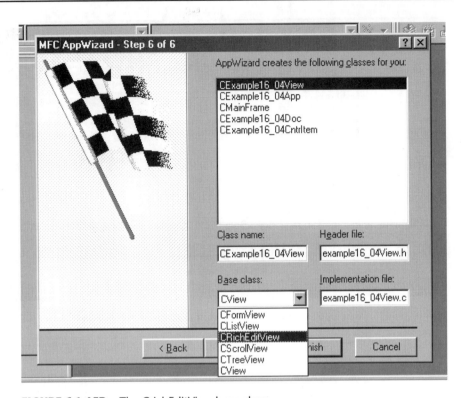

FIGURE 16.45B The CrichEditView base class.

When you are done with the *application wizard*, you will have a base Single Document Interface application (Figure 16.46A). All you will need to do is to fill in the specific code you need.

Now, because you had previously used the `CrichEditView` class as your base class, you can write text, save text files, and open existing text files right in your application! You have just created a basic word processor without writing a single line of code (Figure 16.46B).

This example should illustrate several significant points to you. The first thing you should notice is the sheer power of the *application wizard*. This simple word processor was created for you entirely by the *application wizard*. Secondly, you can see that your choice of base class for your application can be quite important. It was your choice of the `CrichEditView` class that provided the actual support for text files.

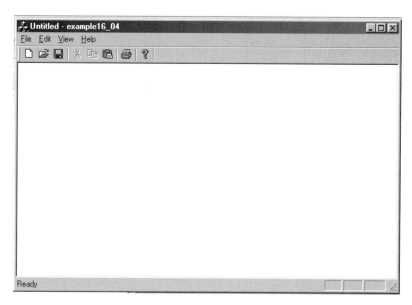

FIGURE 16.46A   The SDI application.

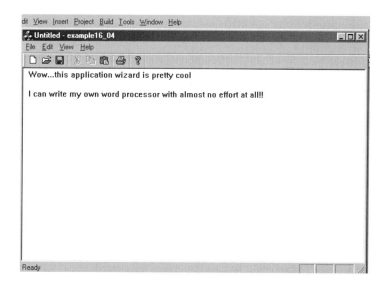

FIGURE 16.46B   The simple word processor

The MDI framework

Much of the MDI application is the same as the SDI application. For that reason, we won't be going over all those details a second time. We will simply highlight the differences.

Step 2: Start the *application wizard*. Follow the *application wizard* as you did in the last example. You will, of course, select the *Multiple Document Interface* option on Step 1 of 6 (Figure 16.47).

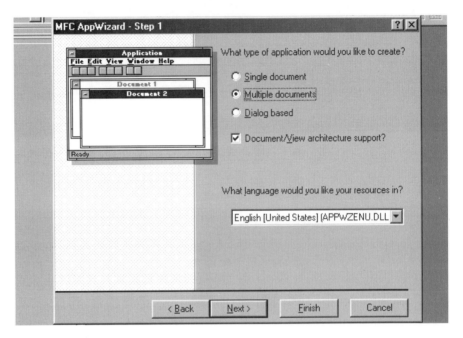

FIGURE 16.47   The MDI application.

You will also choose the container option on Step 3 of 6, just as you did in the SDI application. Keep the default options, just as you did in the SDI application. The real difference will come when you run the application (Figure 16.48). Of course, if you want support for text files, as your SDI application had, you will need to use the CrichEditView as your base class. You will notice that your program is set to display several documents simultaneously. If you will select File and New, a new blank document will be created.

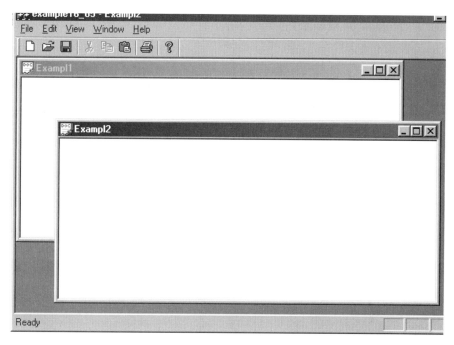

FIGURE 16.48   Running the MDI application

In both of these examples, we simply followed the *application wizard* and created a framework for either a SDI application or a MDI application. These examples are not fully functional, but they do illustrate the ease of setting up the framework of SDI and MDI applications. It is certainly possible to create *Visual C++* applications without the *application wizard*. However, as you have already seen, the *application wizard* creates a number of useful items for you. Essentially, the *application wizard* allows you to focus on your own application-specific code, and not waste time creating toolbars, menus, and the like.

SUMMARY

This chapter took the information presented in Chapter 15 and expanded on it. After these two chapters you should have a solid grasp of how to create simple applications using *Visual C++*. This chapter has shown you how to add menus, new components, and icons to your dialog applications. You now have the tools to add a more professional polish to your applications.

REVIEW QUESTIONS

1. What is the purpose of the radio button?
2. How do you add an existing icon into your program?
3. After typing information into the data tab of the properties window of a combo box, what keys do you press to enter another item?
4. List three methods or properties of the list box.
5. What is a SDI application?
6. What is a MDI application?
7. What is the ID of your dialog's application icon?
8. What class should be used as your base class for SDI and MDI applications if you want to support text documents?

Other Resources

This appendix is provided to allow you to further your studies of any of the items presented in this book. Obviously one book cannot cover it all, so these resources are meant to help you. Each of the resources provided is an excellent place to further your knowledge of a given area. This list does not contain all the available resources for any given topic. Only those resources that the author would personally recommend to a student or a colleague are listed.

C++ COMPILERS

Microsoft *Visual C++*
http://msdn.microsoft.com/visualc/

Borland *C++ Builder*
http://www.borland.com/bcppbuilder/

GCC/GNU
http://gcc.gnu.org/

The *Bloodshed Dev C* Compiler
http://www.bloodshed.net/

C/C++ TUTORIALS

C programming.com
   http://www.cprogramming.com/

History of C
   http://cm.bell-labs.com/cm/cs/who/dmr/chist.html

Stroustrap C++
   http://cm.bell-labs.com/cm/cs/who/dmr/chist.html

Ask the C++ Pro
   http://www.devx.com/gethelp/

C++ CODE SAMPLES AND REFERENCES

Games Programming
   http://www.gamesdomain.com/gamedev/prog.html

Standard Library Reference
   http://www.cplusplus.com/ref/#libs

Data Structures
   http://users.sdsc.edu/~decastro/home/projects/datastruct/

Free source code
   http://www.1cplusplusstreet.com/

Chuck Easttom's *C++ Dungeon*
   http://www.geocities.com/~chuckeasttom/cclass/

      -or-

   http://www.chuckeasttom.com

PROGRAMMING/C++ MAGAZINES

C/C++ Users Journal
   http://www.cuj.com/

Game Developer
   http://www.gdmag.com/

C++ BOOKS

The C++ Programming Language. Bjarne Stroustrap. Addison-Wesley Pub, Co. ISBN: 0201700735

Professional Software Development with Visual C++ 6.0 & MFC. Chao C. Chien. Charles River Media, Inc. ISBN: 1-58450-097-2

C++ *from the Ground up*. Herbert Schildt. Osborne McGraw-Hill. ISBN: 0078824052

C++ *How to Program*. Deitel & Deitel. ISBN 0-13-08971-7
Beginning Visual C++. Ivor Horton. Wrox Press 1-86100-88-X

Glossary of C++ and Programming Terms

#include A mechanism for including one source file into another.

Abstract class A class defining an interface only. These are used as a base class. Declaring a member function pure virtual makes it's class abstract.

Abstraction The act of specifying a general interface hiding implementation details.

Algorithm A systematic way of solving a problem.

Argument A value that is passed to a function, also referred to as a parameter.

Array A variable that consists of a series of values of the same type, the same name, and that are contiguous in memory.

Base Class A class from which another is derived, also called a parent class.

Binary Operators Operators that operate on two arguments.

Bool The built-in Boolean type. It stores true or false values.

Cast Operator for explicit type conversion. This operator allows you to change the data type of a variable.

char Character type; typically a single byte.

cin Standard `istream` operator for reading-in text from the keyboard.

Class A user-defined type. A class can have member functions, member data, member constants, and member types. Classes are templates for creating objects.

Code The actual programming commands that a programmer will write.

Constructor Member function with the same name as its class, used to initialize objects of its class. It is literally used to construct objects based on that class.

cout Standard `ostream` operator used to send output to the default display device, usually the monitor.

Default Constructor Constructor requiring no arguments. Used for default initialization.

Derived Class A class with one or more base classes. Also sometimes called a subclass or child class.

Destructor Member of a class used to clean-up before deleting an object. Its name is its class's name prefixed by the negation symbol ~.

Enumeration A user-defined type consisting of a set of named values

Error Any problem with a program. These can be divided into three categories: logic errors, syntax errors, and runtime errors.

Expression Combination of operators and names producing a value, also called a statement.

Exception An interruption to the normal flow of a program.

FIFO Processing data in a first-in, first-out format, such as in a queue.

float Single-precision floating-point number.

Friend A function or class explicitly granted access to members of a class by that class.

Function A group of one or more related expressions that are grouped under a common name as a discrete code block.

If-Statement Statement selecting between two alternatives based on a condition.

Inheritance The act of getting the public and protected members of a parent class.

LIFO Processing data on a last-in, first-out basis, such as in a stack.

long int Integer of a size greater than or equal to the size of an `int`. Also simply referred to as a `long`.

Member Function A function declared in the scope of a class.

Multiple Inheritance The use of more than one immediate base class for a derived class.

Operator A symbol that defines some operation such as addition, increment, and so on.

Operator Overloading Having more than one operator with the same name in the same scope.

Pointer A variable that actually references the memory address of another variable.

Pure Virtual Function Virtual function that must be overridden in a derived class. A pure virtual function, unlike a normal virtual function, causes the entire class that it is in to be abstract.

Recursion This is the process whereby a function calls itself. This technique is commonly used in a variety of algorithms.

short Integer of a size less than or equal to the size of an `int`.

Statement Combination of operators and names producing some value, also called an expression.

this Pointer to the object for which a non-static member function is called.

Unary Operators Operators that operate on a single argument.

User A person who utilizes your software.

Variable A place in memory set aside to hold data of a particular type.

C

Answers to Review Questions

CHAPTER 1

1. What is a variable?
A place in memory set aside to hold data of a particular type. The variable name is simply a reference to that section of memory.

2. What is a statement?
A line of code that performs some action

3. Which of the following are legal names for variables in C++?
a. `Firstname`   *Yes*
b. `&salary`   *No*
c. `x`   *Yes*
d. `y`   *Yes*
e. `first@name`   *No*

4. All valid C++ statements end with a what?
A semicolon (;)

5. What type of value must the main function return?
An `int`

6. What is the purpose of brackets?
To establish boundaries for a block of code

7. What is a local variable?
A variable declared within some block of code. That variable is only good inside that block of code.

8. What is a global variable?
A variable declared outside any block of code. That means it is good throughout all functions in that file.

9. If a function does not return anything, then what do you put for its return type?
```
void
```

10. What is a parameter?
A variable that is passed to a function

CHAPTER 2
• • • • • • • • •

1. How do you cause the cursor to move to the next line when performing screen output?
Using the \n escape key

2. What does the escape key \b do?
Causes a backspace

3. How would you direct input from the keyboard to a variable called amountentered?
```
cin >> amountentered;
```

4. To use cin, you must import what file?
```
# include <iostream.h>
```

5. Is this statement/expression correct:
```
cout>> "Hello World";
```
No, the arrows (>>) point the wrong way.

6. What does the escape key \t do?
Causes a horizontal tab

7. How would you output a literal text such as "The answer is" and the variable that holds the answer (let's call it answer) on the same line?
```
cout << "the answer is " << answer;
```

8. How would you create a new line if you were writing code in C instead of C++?
The same way as C++. The escape characters work the same in both languages.

9. How would you put literal quotes in your output?
You would use the \" escape character.

10. What is the purpose of the setf function?
It sets the flags for formatting output.

CHAPTER 3
••••••••

1. What is an array?
A series of variables with the same name and the same data type, that are contiguous in memory.

2. What does the ^ operator do?
It does an exclusive bitwise or.

3. What is a multidimensional array?
An array with more than one dimension, such as the array int [5][4] *actually has 20 elements (5 × 4)*

4. A string is actually an _____ of _____.
array, characters

5. List four string methods.
strlen, strcpy, strcat, strncat

6. How would you initialize this array to all 0's int myarray[4];
int myarray[4] = {0,0,0,0};

7. All arrays start at what index number?
0

8. What is the purpose of the cstring header file?
To give you access to the C++ CString *class*

CHAPTER 4
••••••••

1. How do you prototype a function?
Simply place its declaration line at the beginning of your code just after the include statements, and follow it with a semicolon.

2. What are parameters?
Data that you pass to the function when you call it

3. What are the four rules for writing a function?
   a. *The function must have a return type. It may be* void *if the function returns nothing.*
   b. *The function must have a valid name.*
   c. *The function must have parentheses for a list of parameters past to it, even if the parentheses are empty.*
   d. *The function must either be written before it's called, or it must be prototyped.*

4. How many parameters may a function have?
   None, one, or many

5. What is another word for the parameters of a function?
   Arguments

6. What is function overloading?
   Having more than one function with the same name but different types or a different number of parameters.

7. What is one purpose of header files?
   Storing function prototypes

8. What extension do header files end in?
   .h

9. Can an overloaded function have the same number of parameters?
   Yes, if they are of different types.

10. What is passing by reference?
    This is when you pass an address of a variable to a function, rather than the value contained by that variable.

CHAPTER 5
· · · · · · · · ·

1. What are the parts to a for loop declaration?
   • *Declare and initialize the loop counter*
   • *Set the conditions under which the loop will execute*
   • *Increment the loop counter*

2. What types of variables can be used for a switch statement?
   Integer or character

3. What is the purpose of the default statement in a switch statement?
   To handle situations other than the case statements you define. In short, to handle anything that does not fit one of your defined case statements.

4. Why is it important to use the double equals sign in an if statement?
   If you use a single equals, that is an assignment operator, and it will make your if statement always true, regardless of any user input.

5. Why might a switch statement be better than an if statement in some cases?
   If you have multiple selections (more than 2 or 3)

6. What happens if you omit the break statements from your switch block?
   Then the next statement will also execute.

7. What is the purpose of a `for` loop?
   To execute a given code block a certain number of times

8. Can you have an if statement without enclosing brackets?
   Yes, if there is only one line of code and it immediately follows the if statement.

9. How do you implement more than one choice in an if statement?
   With else statements

10. What other possibilities besides equivalence can you use for if statements?
    `<=   >=!= || &&`

CHAPTER 6
• • • • • • • • •

1. What are two modes you can open a file in?
   Binary and Text

2. What is the difference between the `get` method and the `getline` method?
   `Get` *retrieves on character at a time whereas* `getline` *retrieves an entire line at a time.*

3. What header file must you include for file input and output?
   `fstream.h`

4. What is a flat file?
A file with no formatting/relational structure that uses only ASCII codes. Much like a text file created with Microsoft Notepad.

5. What method can you use to change a file's mode?
`setmode`

6. Why would you not want to use the read method to read in plain text?
Because the read *method gives you binary rather than text*

7. What is ASCII?
The American Standard Code for Information Interchange—a 255-character set of numeric values representing the various keys on your keyboard.

8. How does the << operator work with file output?
Much like it does with cout, *it simply outputs the values on the right to the destination on the left.*

9. What is a binary file?
A file that stores the data in a byte-by-byte fashion

10. What is the eof property of a fstream file?
It returns true if you have reached the end of the file in question, false if not.

CHAPTER 7
• • • • • • • •

1. What is the syntax in C++ to trap all errors?
`try{}  catch(...) {}`

2. What are the three general types of errors?
Syntax, logic, runtime

3. What are the four most common mistakes in C++?
Missing semicolon, missing brackets, misspelled words, case sensitivity

4. List three of the basic exception classes in C++.
Runtime_error, exception, invalid_argument, out_of_range. overflow

5. What does it mean to throw an exception?
The exception is passed to the calling function to handle.

6. What happens if you forget to terminate a statement with a semi-colon?
The compiler attempts to execute the next line of code as if it where part of the first statement.

7. What is the purpose of the catch block?
To catch or trap exceptions

8. What is an exception?
An interruption to the normal execution of your program

9. What happens if an exception occurs that is NOT handled by a try-catch block?
The program halts…it crashes.

10. What type of code can you place between the end of the try and the beginning of the catch?
none

CHAPTER 8
• • • • • • • • •

1. What is the purpose of a structure?
To allow you to group related data under a common variable name.

2. Can a structure contain different data types?
Yes

3. What is the reason for using a typedef?
To give a different name to an existing data type

4. Where are structures usually defined?
In a header file

5. How are instances of your structure made?
The same way standard variables are made

6. Can structures be passed as arguments to functions?
Yes

7. A typedef is a type of _____.
variable declaration

8. Can a structure by used as a return type?
Yes

9. What advantage is there to passing a structure by reference?
You can get the data to pass the answers in a single structure.

CHAPTER 9
• • • • • • • • • •

1. What does it mean to say that pointers are type-specific?
   The pointer must be of the same data type as the variable it is pointing to.

2. What happens when you have an int pointer named p and you execute the expression p++?
   The pointer moves over four bytes.

3. What does a pointer point to?
   It points to the address in memory of some variable.

4. Where is one place that pointers are often used?
   With arrays

5. Arithmetic operations on a pointer do what?
   Move the pointer over a certain number of bytes.

6. What is the & operator?
   The address of operator

7. What is **p?
   It is a pointer to a pointer.

8. Why would you want to create a function pointer?
   So you can pass that function as an argument to other functions

9. What is a pointer?
   A variable that points to an address in memory, rather than a value

10. Why don't pointers support multiplication?
    Because pointers contain addresses, multiplication and division do not make sense. If a pointer supported multiplication, the multiplication operation might cause it to point to some memory location that does not exist on your computer.

CHAPTER 10
• • • • • • • • • •

1. What are the four principles of object-oriented theory?
   Abstraction, encapsulation, inheritance, and polymorphism

2. What is encapsulation?
   Placing the data and the functions that work on that data together

3. What is the relationship between structures and classes?
A class is essentially a structure with functions.

4. What is the greatest advantage with object-oriented programming?
Code reusability

5. Why are classes frequently defined in header files?
For two reasons: First, because of their complexity, and, second, to make them more reusable

6. What is inheritance?
When one class gets a copy of the public methods and variables of another class.

7. Give a definition for a class.
A template for creating objects

8. What is the term for a function that is part of a class?
Method

9. What is the term for a variable that is part of a class?
Property

CHAPTER 11
• • • • • • • • •

1. What is inheritance?
It is the process whereby one class gets, or inherits, all the public and protected members of another class.

2. How do you call an overloaded constructor in the base class?
The derived classes constructor is declared to call a specific base class constructor.

3. What is another word for base class?
Parent class

4. How would you declare class a so that it inherits from class b?
`class derived::public class b`

5. What is another term for derived class?
Child class

6. What is a protected variable?
A variable that is private, but can be inherited

7. What is another word for a function that is part of a class?
method

8. Are private members inherited?
No

9. When you have a derived class, which constructor is executed first...the derived class constructor or the base class constructor?
The base class

10. When one class contains another class, what type of relationship is there?
This is a has-a relationship

CHAPTER 12
••••••••••

1. What is polymorphism?
It means functions with the same name but different code

2. What does it mean to override a function?
To use the same name, return type, and parameters, but to change the code in the function

3. How is overriding different from overloading?
In overloading, you change the parameters; in overriding you do not.

4. What is the syntax for a child class to inherit from two base classes (`classa` and `classb`)?
```
childclass: public classa,classb
```

5. What is a function's interface?
Its declaration line (i.e., its return type, name, and parameters)

6. What is a function's implementation
The actual code in the function

7. How do you handle the situation where two base classes have the same function?
Override the function in the derived class

8. What is a virtual function?
A function that must be overridden in any derived class

9. What is an abstract class?
A class that cannot be directly instantiated, but must be inherited to be used

10. How do you create an abstract class?
By making any of its functions a pure virtual function.

CHAPTER 13
• • • • • • • • • •

1. What is a data structure?
   A structured way of storing data, that also defines how that data will be processed

2. Give three examples of data structures.
   Stack, queue, linked list, doubly linked list

3. What is a stack?
   A data structure that stores data on a last-in, first-out basis.

4. What is an algorithm?
   A systematic way of solving a problem

5. What is the basis of the bubble sort?
   To compare two adjacent elements in an array or list and to switch them if they are not in order

6. What two methods must be associated with a linked list?
   Push and pop.

7. What is a doubly linked list?
   A linked list that has references to the item before it in the list and the item after it in the list

8. What does the word static do?
   It causes a local variable's value to be retained between iterations of a function

9. What is recursion?
   It is the process whereby a function calls itself

CHAPTER 14
• • • • • • • • • •

1. Why are loop structures so important for games?
   They allow you to keep repeating the game until the user elects to stop the game. They are also frequently used in creating graphics.

2. What is *Direct X*?
   A multimedia library for Windows programmers

3. What is the real key to games programming?
   Creative application of standard programming techniques

4. Why are if statements so important for games programming?
You must take different courses of action depending on the user's choice of actions.

5. What are some uses of classes in games?
They can be used to represent objects within the game.

6. What are ASCII-based graphics?
Graphics created using simple keyboard-generated symbols, such as asterisks.

CHAPTER 15
• • • • • • • • • •

1. What is the MFC?
Microsoft Foundation Classes, a set of 200+ classes that are wrappers for Windows API functions.

2. What method is used to move data from variables to the associated components?
`UpdateData(TRUE)`

3. List five common components used in *Visual C++*.
`Static Box, Edit Box, Picture, Combo Box, List Box, Button, Tab Control, Tree Control, List Control`

4. List 4 methods of `CString`.
`GetLength, MakeUpper, MakeLower, MakeReverse, find, compare mid`

5. What two functions do all *Windows* programs have?
`WinMain` *and* `WinProc`

6. What header file must be included in order to use the `message box` function?
`stdafx.h`

7. What is the purpose of the *Class Wizard*?
To automate the process of mapping messages to functions, and mapping variables to components

8. What is the title of the second tab in the *Class Wizard*?
`Member Variables`

9. List at least two constants for the `message beep` function.
`MB_ICONASTERISK, MB_ICONHAND, MB_OK, MB_ICONEXCLAMATION`

CHAPTER 16
• • • • • • • • • •

1. What is the purpose of the `radio button`?
   To allow the user to select from multiple mutually exclusive options

2. How do you add an existing icon into your program?
   By right-clicking on the icon heading in resources

3. After typing information into the data tab of the properties window of a `combo box`, what keys do you press to enter another item?
   `CTRL-ENTER`

4. List three methods or properties of the `list box`.
   `Sort, border, selection`

5. What is a SDI application?
   Single Document Interface

6. What is a MDI application?
   Multi Document Interface

7. What is the ID of your dialog's application icon?
   `IDR_MAINFRAME`

8. What class should be used as your base class for SDI and MDI applications if you want to support text documents?
   CrichEditView

The C++
Builder Compiler

B orland sells a wide range of development tools. These tools are quite useful. However, many people, especially students just learning about programming, need something cheaper (preferably free!) to work with. Borland has been kind enough to provide a free command line compiler. It is the one used throughout most of this book (except for the chapters on *Visual C++*). This appendix will walk you through the download, installation, configuration, and use of this free compiler.

Step 1: The first step is to download the compiler. First, you will need to go the Borland Web site, where there is an entire page devoted to this free compiler.

http://www.borland.com/bcppbuilder/freecompiler/

You can simply click on "Try it Now" on the lefthand side of the screen, just as you see in Figure D.1.

Once you select that option you will be prompted to register with Borland, and then to download the compiler to your machine. What you actually download is a setup file that you will need to run. Figure D.2 is a picture of what it would look like if viewed in "My Computer" on a *Windows* PC.

Once you launch the application, it will ask you where you wish to install the product. This will look much like the image in Figure D.3.

FIGURE D.1   Downloading the Borland compiler.

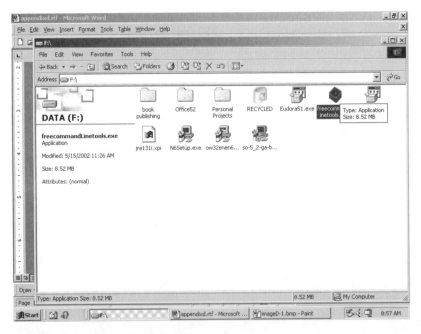

FIGURE D.2   The Borland setup file.

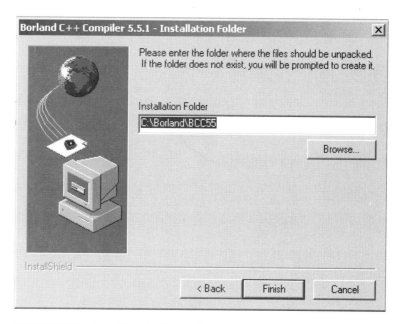

FIGURE D.3   Installing the Borland compiler.

Once the installation is done, you will have to configure the compiler. In whatever directory you install this compiler to you will notice the following items:

1. A `license.txt`
2. A `readme.txt`
3. A `Bin` directory
4. An `Examples` directory
5. A `Help` directory
6. A `Lib` directory
7. An `Include` directory

Please note that, although this is a good compiler, the `readme` file is of almost no help at all. The first step you will have to do is to create two text files called `bcc32.cfg` and `ilink32.cfg`. These are configuration files that tell your compiler where to find things. To create them, open your favorite text editor and enter the following.

```
In bcc32.cfg
I"c:\Borland\Bcc55\include"
L"c:\Borland\Bcc55\lib;
```

What these two lines will do is to tell the compiler where it can find `include` files, and libraries.

```
In ilink32.cfg
L"C:\Borland\Bcc55\lib;
```

This line tells the linker where to find libraries that need to be linked-in when your program is compiled.

HINT!

If you installed this to a different directory, then you will have to put the path to the following directory: `c:\Borland\Bcc55\`

You are almost ready. Now you just need to add this path to your system's path variables. Path variables allow your operating system to know where to find things. This is done different ways for different operating systems. If you have *Windows 2000,* then you will go to `Start > Control Panel > System`, and choose the `Advanced Tab`. Next, click the `Environment Variables` button. Then select `Path`. This should look much like what you see in Figure D.4.

When you click the `Edit` button, you will see a path statement. At the very end, simply place the following line.

```
c:\Borland\BCC55\Bin.
```

This should resemble the image you see in Figure D.5.

HINT!

If you installed to a different directory, then please use that directory path.

Now you are set, and ready to compile!

If, however, you are using an older version of *Windows,* you can go to the DOS prompt and type-in the following.

```
Set path = %path%;c:\Borland\BCC55\Bin;
```

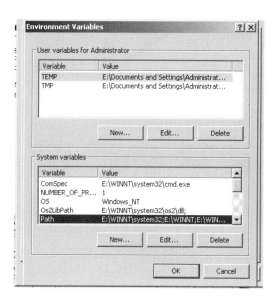

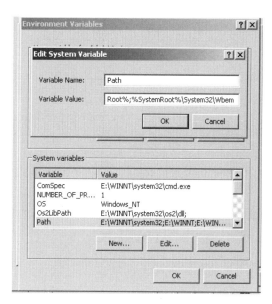

FIGURE D.4 Setting environmental variables.   **FIGURE D.5** Setting the path statement.

For any operating system, you can simply consult the documentation of that operating system and find out how to set the path statement. What you are doing, whatever process is required, is letting the operating system know where the compiler is located, so that you can invoke it from anywhere on your PC. Now you will be able to compile your C++ programs. Simply save your code in a plain text file that ends with a *.cpp* extension. Then, from the command line, type-in the following.

```
bcc32 filename.cpp
```

When that is completed, if there are no errors in your code, you will then have an executable you can distribute: `filename.cpp`

If, however, you have errors, then you will need to correct those before you will be able to proceed.

If, for any reason, setting the path statement does not work, you can get around this by copying the source file you wish to compile to the folder where your compiler is:

```
C:\Borland\bcc55\bin\
```

Then you should be able to compile it.

E

Common
Mistakes

This appendix is provided to show you a number of mistakes that beginners commonly make. Hopefully, if you review this you will avoid some of these mistakes. At the very least, you may want to consult this appendix if you are trying to find some bug in your program.

THE MISSING SEMICOLON

All statements/expressions in C++ (as well as in *Java* and C) end with a semicolon. If you leave that off, then the compiler tries to interpret the next line of code as being part of the line that is missing the semicolon. This can lead to a number of very strange error messages when you attempt to compile.

THE EXTRA SEMICOLON

You only need semicolons at the end of statements. You do NOT place them after any of the following:

1. Function declarations
2. Include statements

3. `If` statements
4. `Switch` statements

THE SINGLE EQUALS

When using an `if` statement, remember that the single equals is an assignment operator; the double equals is an evaluation operator. Make sure you use the double equals with `if` statements.

WRONG NUMBER OF BRACKETS

Remember that brackets are barriers around blocks of code. They show where the block of code starts and where it stops. You must have a closing bracket for each and every opening bracket. `If` you have too many or too few closing brackets, then the compiler will try to make sense of your code, but will generate some very peculiar errors. It is usually a good idea to check your code to make sure you have one and only one closing bracket for each and every opening bracket.

MISSPELLINGS

It is very common for beginning programmers to forget to check the spelling of their variables.

CAPITALIZATION

Remember that C++ is case-sensitive. `int Main()` is not the same as `int main()`!

INITIALIZING POINTERS

You cannot use a pointer until you have initialized it. Failure to initialize a pointer before using it will cause an error.

OVERFLOWING AN ARRAY

If you attempt to access elements in an array, beyond the range you declared, this can lead to serious problems. You must make certain that all array operations are taking place within the bounds of the array. If you declare an array as `int myarray[10]`, you must make sure you only attempt to access elements 0 through 9.

Overflowing the array is most likely to occur when using a pointer to an array, and then using an increment operator with the pointer. It is very easy to overlook the bounds of your array and find yourself pointing to some memory address that is outside the bounds of your array.

IMPROPER FOR LOOP DECLARATIONS

When you create a `for` loop, the second element of the declaration tells the loop under what conditions to execute. It does not tell the loop where to stop. For example if you say the following:

```
for (int j = 0; j < 10; j++)
```

You have told the loop to execute as long as `j` is less than 10. This is exactly correct. A common beginner mistake is to write this as:

```
For (int j = 0; j = 10;j++)
```

This will never work. You have instructed the loop to execute as long as `j` is equal to 10, yet you initialized `j` to 0. That means the loop will never execute!

MISSING *VISUAL C++ APP WIZARD STEPS*

After a person has gotten comfortable with the *Application Wizard*, it is common to rush through it. This can lead to many problems later on. If, for example, in your SDI and MDI applications, you fail to use `CrichEditView` as your base class, then your application will not actually support text documents. This is done during Step 6 of the *Application Wizard* for SDI and MDI applications.

About the CD-ROM

The accompanying CD-ROM is given to you so that you can have access to fully working, code samples. There are over 80 fully working examples in the book and on the CD-ROM. Each chapter of the book has an accompanying folder on the CD-ROM. In that folder you will find each and every one of the examples from that chapter, ready for you to compile and run. The following table shows a sample of the folders and their contents:

Folder	Contents
Chapter 02	02-01.cpp
	02-02.cpp
	02-03.cpp
	02-04.cpp
	02-05.cpp
	02-06.cpp

Folder	Contents
Chapter 03	03-01.cpp
	03-02.cpp
	03-03.cpp
	03-04.cpp
	03-05.cpp
	03-06.cpp
	03-07.cpp
	03-08.cpp
	03-09.cpp
Chapter 04	04-01.cpp
	04-02.cpp
	04-03.cpp
	04-04.cpp
	04-05.cpp
Chapter 05	05-01.cpp
	05-02.cpp
	05-03.cpp
	05-03b.cpp
	05-03.h
	05-04.cpp
	05-04b.cpp
	05-05.cpp
Chapter 06	06-01.cpp
	06-02.cpp
	06-03.cpp
	06-04.cpp
Chapter 07	07-01.cpp
	07-02.cpp
	07-03.cpp

Folder	Contents
Chapter 08	08-01.cpp
	08-02.cpp
	08-03.cpp
	08-04.cpp
	08-05.cpp
	08-06.cpp
Chapter 09	09-01.cpp
	09-02.cpp
	09-03.cpp
	09-04.cpp
	09-05.cpp
Chapter 10	10-01.cpp
	10-02.cpp
	10-02.h
	10-03.cpp
	10-04.cpp
	10-05.cpp
	10-06.cpp
	10-07.cpp
Chapter 11	11-01.cpp
	11-01.h
	11-02.cpp
	11-03.cpp
	11-03.h
	11-04.cpp
	11-04.h
	11-05.cpp
	11-05.h
	11-06.cpp
	11-06.h

Folder	Contents
Chapter 12	12-01.cpp
	12-01.h
	12-02.cpp
	12-02.h
	12-03.cpp
	12-03.h
	12-04.cpp
	12-04.h
Chapter 13	13-01.cpp
	13-02.cpp
	13-03.cpp
	13-04.cpp
	13-05.cpp
	13-06.cpp
	linkedlist.h
Chapter 14	14-01.cpp
	14-01.h
	14-02.cpp
	14-03.cpp
Chapter 15	Subdirectories
	Example15_01
	Example15_02
	Example15_03
	Example15_04
Chapter 16	Subdirectories
	Example16_01
	Example16_02
	Example16_03
	Example16_04
	Example16_05

Hopefully these code samples will help you in your learning of C++. If you have trouble making any example in the book work, check the code you have written against the matching source file on the CD-ROM. These files are standard C++ source files that can be opened and compiled by any standard C++ compiler (Microsoft *Visual C++*, Borland *C++ Builder*, *Visual Age C++*, and many of the free command line compilers available on the Internet).

SYSTEM REQUIREMENTS

To use the files on the book's CD-ROM, you need to meet the following requirements:

- PC: Windows 98, 2000, NT, ME, XP or Linux
- 200 MHz or faster; 32 MB RAM; 20 MB hard drive space;
- CD-ROM drive; VGA monitor; Borland C++ compiler

For *Visual C++* Section:

- Windows 98, 2000, NT, ME, XP;
- 200 MHz or faster; 32 MB RAM; 20 MB hard drive space;
- CD-ROM drive; VGA monitor; *Visual C++* compiler.

Index